Practice Papers for SQA Exams

Higher

Human Biology

D1355012

Introduction

Layout of the book

The four papers that follow have been produced to give you practice in two of the elements needed for Higher Grade Human Biology, reflecting Curriculum for Excellence's values and purposes. The first element is 'Knowledge and Understanding' [KU], which requires you to be familiar with all the theory. This is especially important in Human Biology, which is very much a knowledge-based subject: you need to know the facts, concepts and common techniques. The second element is 'Problem-Solving' [PS], which tests your ability to: select relevant information from a variety of sources; present information in a variety of forms; process information; plan, design and evaluate the design of experiments; draw valid conclusions and give explanations; and make predictions and generalisations. These two elements, which are assessed by the formal national exam are combined to produce your overall grade. As a very rough guide, you will be awarded a pass at A, B, C if you obtain around 75%, 65% and 55% respectively.

All four papers have been designed to mirror the actual national exam. The layout and level of difficulty of all four papers are very closely modelled on the national exam you will sit, so you will become familiar with how it looks and how best to tackle it. In addition, the format of the questions reflects what you are likely to encounter in the exam. All parts of the syllabus have been represented in these papers.

Each paper has two sections:

- Section 1 consists of 20 multiple-choice questions, which test KU and PS. This section is worth 20 marks and is answered on a separate grid that is provided for the exam.

- Section 2 comprises restricted and extended response questions with a mixture of KU and PS. The majority of marks are for applying knowledge and understanding, with the other marks for applying scientific inquiry, scientific analytical thinking and problem-solving skills. This section is worth 80 marks and the questions are answered in spaces on the exam paper.

- The whole paper is worth 100 marks and you will have 2 hours and 30 minutes to complete it.

Pages 8&9 of this book have a **Topic index**, which help you locate questions linked to particular sections of the syllabus or to one of the seven skills specified by the Scottish Qualifications Authority.

The answer section is at the back of the book. Each answer contains, where appropriate, some guidance as to how this was obtained, practical tips on how to tackle certain types of questions, details of how marks are awarded and advice on just what the examiners will be looking for. Also included is the type of question: knowledge and understanding or problem-solving.

How to use this book

Interacting with learning material is a powerful way to obtain feedback on your strengths and weaknesses. The material in these papers will give you an excellent way of working on different strategies needed to handle the Higher exam. Seeking help – where needed – from your teacher is vital to improvement and building up your confidence and expertise.

The papers can be used in two main ways:

1. You can complete an entire paper as preparation for the final exam. If you would like to use the book in this way, you can complete the paper under exam-style conditions by setting yourself a time for each paper and answering it as well as possible without using any references or notes. Alternatively, you can answer the paper questions as a revision exercise, using your notes to produce a model answer. Your teacher may mark these for you.

2. You can use the **Topic index** section on pages 8&9 to find all the questions within the book that deal with a specific topic in any of the four Course Units. This allows you to focus specifically on areas that you particularly want to revise or, if you are mid-way through your course, it lets you practise answering exam-style questions for just those topics that you have studied. You will find this particularly useful if you want practice on handling particular types of problem-solving questions.

Revision advice

The need to work out a plan for regular and methodical revision is obvious. If you leave things to the last minute, it may result in panic and stress which will inhibit you from performing to your maximum ability. If you need help, it is best to find this out when there is time to put it right. Revision planners are highly individual and you need to produce one that suits you. Use an area of your home that is set aside only for studying if possible, so that you form a positive link and in this way will be less liable to distractions. Not only do you need a plan for revising Higher Human Biology, but also for all your subjects. Below is one revision plan but you will have your own ideas here!

Work out a revision timetable for each week's work in advance – remember to cover all of your subjects and to leave time for homework and breaks. For example:

Day	6.00–6.45 pm	7.00–8.00 pm	8.15–9.00 pm	9.15–10.00 pm
Monday	Homework	Homework	English revision	Human Biology revision
Tuesday	Maths revision	Physics revision	Homework	Free
Wednesday	Geography revision	English revision	Human Biology revision	Maths revision
Thursday	Homework	Physics revision	Geography revision	Free
Friday	English revision	Human Biology revision	Free	Free
Saturday	Free	Free	Free	Free
Sunday	Maths revision	Physics revision	Geography revision	Homework

Make sure that you have at least one evening free each week to relax, socialise and re-charge your batteries. It also gives your brain a chance to process the information that you have been feeding it all week.

Arrange your study time into sessions that suit you, with a 15-minute break in between. Try to start studying as early as possible in the evening, when your brain is still alert, and be aware that the longer you put off starting, the harder it will be.

If you miss a session, do not panic. Log this and make it up as soon as possible. Do not get behind in your schedule – discipline is everything in being a successful student.

Study a different subject in each session, except for the day before an exam.

Do something different during your breaks between study sessions – have a cup of tea, or listen to some music. Do not let your 15 minutes expand into 20 or 25 minutes!

Have your class notes and any textbooks available for your revision to hand, as well as plenty of blank paper, a pen, etc. If relevant, you may wish to have access to the Internet but be careful you restrict using this only for supporting your revision. You may also like to make keyword sheets like the example below:

Keyword	Meaning
Ribosome	Structure in the cell that manufactures protein
Fungicide	Chemical that kills fungi

Flash cards are another excellent way of practising terms and definitions. You can make these easily or buy them very cheaply. Use flash cards either to recall the keyword when you see the meaning, or to give the meaning when you see the keyword. There are several websites that are free to use and give you the ability to generate flash cards online. If you collaborate with your friends and take different sections of the course, you can merge these into a very powerful learning and revision aid.

Finally, forget or ignore all or some of the advice in this section if you are happy with your present way of studying. Everyone revises differently, so find a way that works for you!

Command words

In the papers and in the Higher exam itself, a number of command words will be used in the questions. These command words are used to show you how you should answer a question: some words indicate that you should write more than others. If you familiarise yourself with these command words, it will help you to structure your answers more effectively.

Command word	Meaning/explanation
Name, state, identify, list	Giving a list is acceptable here – as a general rule you will get one mark for each point you give.
Suggest	Give more than a list – perhaps a proposal or an idea.
Outline	Give a brief description or overview of what you are talking about.
Describe	Give more detail than you would in an outline, and use examples where you can.
Explain	Discuss why an action has been taken or an outcome reached – what are the reasons and/or processes behind it?
Justify	Give reasons for your answer, stating why you have taken an action or reached a particular conclusion.
Define	Give the meaning of the term.
Compare/contrast	Give the key features of **two** different items or ideas and discuss their similarities/differences.
Predict	Work out what will happen.

In the exam

Watch your time and pace yourself carefully. Work out roughly how much time you can spend on each answer and try to stick to this.

Be clear before the examination what the instructions are likely to be, for example how many questions you should answer in each section. The papers will help you to become familiar with the examination instructions.

Read the question thoroughly before you begin to answer it – make sure you know exactly what the question is asking you to do. If the question is in sections, for example, question 12 on page 64, make sure that you can answer each section before you start writing.

Plan your extended responses by jotting down keywords, making a brief mind-map of the important points or whatever you find works best for you.

Do not repeat yourself as you will not get any more marks for saying the same thing twice. This also applies to annotated diagrams, which will not get you any extra marks if the information is repeated in the written part of your answer.

Give proper explanations. A common error is to give descriptions rather than explanations. If you are asked to explain something, you should be giving reasons. Check your answer to an **explain** question, and make sure that you have used linking words and phrases such as **because, this means that, therefore, so, so that, due to, since** and **the reason is**.

Good luck!

Paper brief and question allocation by type

The paper brief on page 7 shows the breakdown of questions within the paper.

This will enable you to allocate your revision proportionately to maximise your chance of success. For example, with KU questions making up approximately 70 of the marks, you should spend approximately 70% of your time revising your notes and doing practice questions on KU.

KU is split into two areas: **demonstrating** knowledge and **applying** knowledge.

- Demonstrating involves mostly recall, whereas applying involves you using that knowledge in unfamiliar situations that you won't have learned.

- The best way to improve the application of knowledge is to practise from papers and textbooks.

- In these questions you will also be provided with context and it is important to read all the information provided in the question to allow you to identify the knowledge you need to recall or apply.

The **skills section**, approximately 30 marks, involves data handling and experimental design questions. The skills revision should take up approximately 30% of your revision time.

- The **data handling** questions, approximately 5–9 marks, give you the data to process or make predictions or conclusions. The best way to improve your technique with these questions is practice. The past papers and textbooks will provide you with the opportunity to do this. Be sure to read all of the data and information carefully and, as always, read the question more than once. You are often asked to use only part of the data to answer a particular question.

- The **experimental design** questions, approximately 5–9 marks, give you lots of information about an experiment and you will be asked questions about the design or results provided. Once again, the key to answering successfully is to read all of the data and information carefully and to do as many practice questions as you can. The questions asked are often similar, but it is the experiment which differs.

The table below details where these areas are covered within these papers, so if you are struggling with one area, you can do multiple questions to practise.

Component	Marks		
	Knowledge and Understanding	Skills	Total
Question Paper	70±5	30±5	100

Knowledge and Understanding	Marks
Demonstrating knowledge and understanding of human biology by making statements, describing information, providing explanations and integrating knowledge.	min 25
Applying knowledge and understanding of human biology to new situations, interpreting information and solving problems.	min 25

Skills	Marks
Planning and designing experiments/investigations.	
Selecting information from a variety of sources.	
Presenting information appropriately in a variety of forms.	
Processing information/data (using calculations and units, where appropriate).	25–35
Making **predictions** and generalisations based on evidence/information.	
Drawing valid **conclusions** and giving explanations supported by evidence/justification.	
Evaluating experiments/investigations and suggesting improvements.	

Additional information

Question type	Question number			
	Paper A	Paper B	Paper C	Paper D
Two or three extended writing questions: 10–15 marks in total. At least one of the extended writing questions will include a choice of topic.	12 A 12 B	12 A 12 B	14 A 14 B	13 A 13 B
One large data handling question: 5–9 marks.	10	4	10	6
One large experimental design question: 5–9 marks.	4	7	3	2
'A' type marks: questions set at grade A level of difficulty – approximately 30%.	Throughout paper, usually with keywords like 'explain' or 'describe' and worth more than 1 mark.			

Topic Index

Skill Tested	Key Area	Paper A S1	Paper A S2	Paper B S1	Paper B S2	Paper C S1	Paper C S2	Paper D S1	Paper D S2
Unit 1: Human Cells — Demonstrating and applying knowledge	1. Division and differentiation in human cells	1	1				1	1	
	2. Structure and replication of DNA	4			1	1			1
	3. Gene expression	5	2 3(d)	1		2		3	
	4. Genes and proteins in health and disease		8(d) (e)	2		3	2(a)	4	
	5. Human genomics	6			2	4			3(a) (b) (d) (f)
	6. Metabolic pathways		3(a) (b)	4		5		6	
	7. Cellular respiration		3(c)	6			4 5	7	
	8. Energy systems in muscle cells	7			3		6		
Unit 2: Physiology and Health — Demonstrating and applying knowledge	1. Structure and function of reproductive organs and gametes and their role in fertilisation			7		9	14A		4(a) (i) (iv)
	2. Hormonal control of reproduction	8	12A			10		8	4(a) (ii) (iii) (iv)
	3. The biology of controlling fertilisation			8	5(c) (d) (e)				4(b) 13A
	4. Ante- and postnatal screening		8(a) (b) (g) (h)	9			2(b) (c) (d) 8 9	10	5(a) (d)
	5. Structure and function of arteries, capillaries and veins	10	5			11		11	
	6. Structure and function of the heart		11(c)	10	8(d) 12B			13	
	7. Pathology of cardio-vascular disease	11	6	11		12			
	8. Blood sugar levels and obesity		9(c)		4 (b)	14			

Skill Tested	Key Area	Paper A		Paper B		Paper C		Paper D	
		S1	S2	S1	S2	S1	S2	S1	S2
Unit 3 : Neurobiology and Communications — Demonstrating and applying knowledge	1. Divisions of the nervous system and parts of the brain	14				15	11 12(a)	18	8
	2. Perception and memory	15	7(a) (b) (c)	14	9(g) (h)			15	
	3. Cells of the nervous system and neurotransmitters at synapses			15	12A				9
	4. Communication and social behaviour	16	10(f) (g)	16		17	13	19	
Unit 4 : Immunology and Public Health — Demonstrating and applying knowledge	1. Non-specific defences			17		18		20	13B (i)
	2. Specific cellular defences	17			11				11
	3. Transmission and control of infectious diseases	18	12B						12
	4. Active immunisation and vaccination and the evasion of specific immune response by pathogens	19		18	6(b) 10		14B		13B (ii)
Higher Human Biology Course — Skills of scientific enquiry	Selecting information			12	7(b) 8(a)	19		9 16	2(a) 3(c) 6(a) (ii) (iii) (iv) 6(b)
	Presenting information		4(c)		7(g) 9(a)		3(b)		2(d)
	Processing information	2 3 9 12	8(c) (f) 9(a) (b) 10(a) 11(a) (b)	3 5	4(c) 5(b) 7(a) 8(b) 9(d)	6 7 8 13	3(a) (c) (v) 7 10(a) (b) (c) (d)	2	3(e) 5(b) (c)
	Planning investigations		4(a) (c) (d) 10(c)	13	7(d) (e)				2(c) 7(b)
	Evaluating experimental procedures		4(b) 7(d) 10(d) (e)		6(a) 7(f) 8(c) 9(e) (f)		3(c) (i) (ii) (iii)		10(b) (c)
	Drawing conclusions	13 20	4(f) 10(b)	19 20	7(h)	16	10(e) 3(c) (iv)	5 14	2(e) (6)(a) (i) 7(a) 10(a)
	Predicting and generalising				4(a) (d) 5(a) 7(c) 9(b) (c)	20	10(f) 12(b)	12 17	2(b)

Paper A

Higher Human Biology

Practice Papers for SQA Exams

Paper A

You are allowed 2 hours and 30 minutes to complete this paper.

A total of 100 marks are available in this paper.

Section 1 – 20 marks

Section 2 – 80 marks

Attempt ALL questions.

Scotland's leading educational publishers

SECTION 1 – 20 marks

Attempt ALL questions.

Answers should be given on the separate answer sheet provided.

1. Which of the following correctly identifies **three** main body tissue types?

   A   Epithelial, connective, muscle

   B   Nerve, epithelial, skeletal

   C   Connective, muscle, skin

   D   Nerve, muscle, heart.

2. One form of cancer treatment involves chemotherapy which uses a drug to kill dividing cells.

   The graph shows the number of healthy cells and cancer cells in the blood of a patient receiving chemotherapy. The arrows labelled F to I show when the drug was given to the patient.

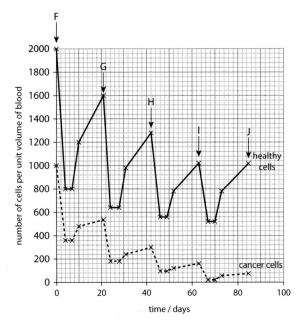

   Calculate the simplest whole number ratio of healthy cells to cancer cells, per unit volume of blood, at day 4.

   A   2:1

   B   20:9

   C   80:36

   D   400:180

3. If 20% of the bases in a molecule of DNA are thymine, what is the ratio of thymine to cytosine in the same molecule?

   A   1:2

   B   2:3

   C   3:4

   D   4:5

4. Which line in the table correctly describes the structure of DNA?

	Nucleotide contains	Backbone composed of
A	deoxyribose sugar, phosphate and base	hydrogen bonds
B	ribose sugar, phosphate and base	hydrogen bonds
C	deoxyribose sugar, phosphate and base	sugar–phosphate
D	ribose sugar, phosphate and base	sugar–phosphate

5. Exons in the primary transcript of mRNA are

   A   non-coding regions and are joined together to form a mature transcript

   B   coding regions and are joined together to form a mature transcript

   C   non-coding regions and are removed in RNA splicing

   D   coding regions and are removed in RNA splicing.

6. After PCR, what is used to allow detection of DNA sequences?

   A   Repeated cycles of heating and cooling

   B   Addition of heat-tolerant DNA polymerase

   C   Use of complementary primers

   D   Fluorescent labelling.

7. Which line in the table below correctly describes type 1 muscle fibres?

	Contraction speed	Contraction duration
A	slowly	longer
B	quickly	longer
C	slowly	shorter
D	quickly	shorter

8. Which of the following statements correctly describes the role of follicle-stimulating hormone (FSH) in males?

A   Stimulates the production of testosterone

B   Promotes sperm production

C   Activates the prostate gland and seminal vesicles

D   Stimulates the pituitary gland.

9. The sperm counts of a sample of men taken between 1950 and 2010 are shown in the graph below.

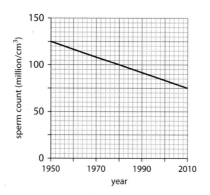

What is the average reduction in sperm count per year?

A   0·75 million/cm³/year

B   0·83 million/cm³/year

C   0·92 million/cm³/year

D   50 million/cm³/year

10. The diagram below shows a blood vessel.

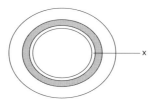

Which of the following is structure X?

A   Endothelium

B   Central lumen

C   Elastic fibre

D   Smooth muscle.

11. Which of the following correctly states the effect of exercise on reducing the risk of developing cardiovascular disease (CVD)?

   A   Keeping weight under control, minimising stress, reducing hypertension and improving HDL blood profiles

   B   Keeping weight under control, increasing stress, increasing hypertension and improving HDL blood profiles

   C   Losing lots of weight, minimising stress, reducing hypertension and increasing HDL blood profiles

   D   Keeping weight under control, minimising stress, reducing hypertension and increasing HDL blood profiles.

12. The cell shown below is magnified five hundred times. What is the actual size of the cell?

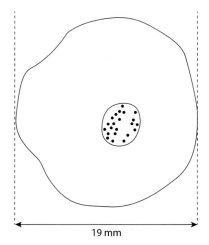

19 mm

   A   1900 μm

   B   95 μm

   C   38 μm

   D   3·8 μm

13. People with type 2 diabetes have cells with low sensitivity to insulin. About 80% of people with type 2 diabetes are overweight or obese. Some people who are obese have gastric bypass surgery (GBS) to help them to lose weight.

Doctors investigated whether GBS affected sensitivity to insulin. They measured patients' sensitivity to insulin before and after GBS. About half of the patients had type 2 diabetes. The other half did not but were considered at high risk of developing the condition.

The table shows the doctors' results. The higher the value in the table, the greater the sensitivity to insulin.

Patients	Mean sensitivity to insulin/arbitrary units	
	Before gastric bypass surgery	1 month after gastric bypass surgery
Did not have diabetes	0·55	1·30
Had type 2 diabetes	0·40	1·10

Which of the following conclusions can be drawn from the data **in the table**?

A   People with type 2 diabetes have a greater sensitivity to insulin after GBS than those who do not have type 2 diabetes

B   People with type 2 diabetes have a decreased sensitivity to insulin after GBS compared with those who do not have type 2 diabetes

C   People with type 2 diabetes are more likely to have GBS than those who do not have type 2 diabetes

D   People with type 2 diabetes are more obese than those who do not have type 2 diabetes.

14. Which of the following does **not** correctly describe the function of the cerebral cortex?

A   Receiving sensory information

B   Co-ordinating voluntary movement

C   Making decisions in the light of experience

D   Regulating the basic life processes of breathing, heart rate, arousal and sleep.

15. The diagram below shows the three levels of memory, X, Y and Z.

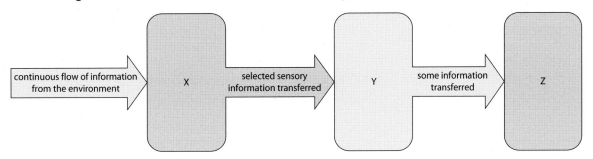

Select the correct type of memory for X, Y and Z.

	X	Y	Z
A	sensory	long-term	short-term
B	sensory	short-term	long-term
C	short-term	long-term	sensory
D	short-term	sensory	long-term

16. Which of the following terms is used to describe a loss of personal identity in a group leading to diminished restraints on behaviour?

A   Social facilitation

B   Internalisation

C   Deindividuation

D   Identification.

17. Which of the following correctly describes possible actions of T and B lymphocytes?

	T lymphocytes	B lymphocytes
A	destroy infected cells by inducing apoptosis	attack insulin producing cells in type 1 diabetes (auto-immune response)
B	secrete antibodies into the lymph and blood where they make their way to the infected area	activated by antigen presenting cells
C	secrete cytokines	attack insulin producing cells in type 1 diabetes (auto-immune response)
D	may become memory cells	secrete antibodies into the lymph and blood where they make their way to the infected area

18. Which of the following terms would describe the spread of a disease which occurred occasionally?

A   Sporadic

B   Endemic

C   Epidemic

D   Pandemic.

19. Which one of the following diseases has a pathogen able to survive within phagocytes and avoid immune detection?

A   HIV

B   Influenza

C   Tuberculosis

D   Malaria.

20. Nutritionists investigated the relationship between eating oily and non-oily fish and the incidence of asthma. They analysed the diets of children with asthma and the diets of children without asthma.

The figure shows the results from their investigation.

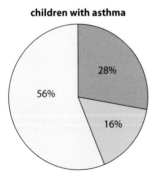

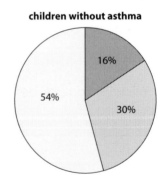

key
- children who ate no fish
- children who ate oily fish
- children who ate non-oily fish

Which of the following conclusions is **not** correct, based only on the data **in the figure**?

A   It is less likely that children with asthma ate oily fish

B   It is less likely that children with asthma ate fish

C   There is only 2% difference in children with or without asthma who ate non-oily fish

D   28% of children without asthma ate oily fish.

SECTION 2 – 80 marks

Attempt ALL questions.

It should be noted that question 12 contains a choice.

1. The diagram below shows a somatic cell undergoing mitosis.

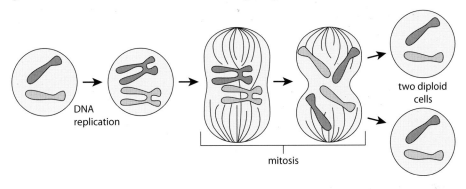

(a) State another type of cell which divides by mitosis but also by meiosis. **1**

_____

(b) Stem cells are unspecialised somatic cells that can differentiate into specialised cells.

Name **two** types of stem cells. **1**

_____

(c) Give **one** difference between these two types of stem cells. **1**

_____

_____

(d) Describe what is meant by cellular differentiation. **1**

_____

_____

1. (continued)

(e) Give an account of the research and therapeutic uses of stem cells and the issues surrounding their use.

3

_____

_____

_____

_____

_____

(f) Cancer cells divide excessively to produce a mass of abnormal cells called a tumour.

Describe what can happen if cancer cells fail to attach to each other.

1

_____

_____

2. The diagram below shows the process of transcription during gene expression.

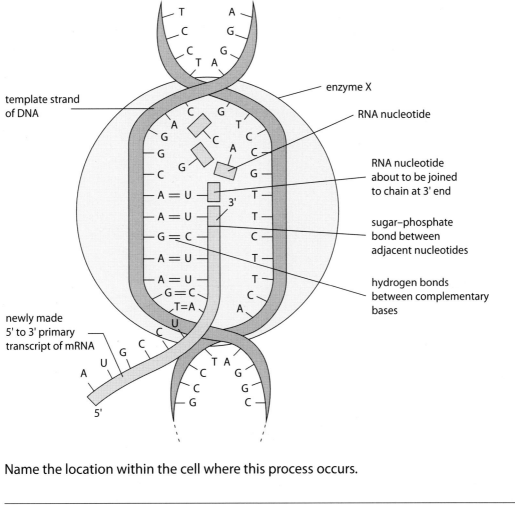

(a) Name the location within the cell where this process occurs. **1**

_____

(b) Name enzyme X. **1**

_____

(c) Describe the function of this enzyme. **2**

_____

_____

_____

_____

2. (continued)

(d) The diagram shows that RNA has the base uracil, while DNA has the base thymine.

State **one** other difference between RNA and DNA. **1**

_____

_____

(e) The primary transcript of RNA can then undergo a process.

State the name of this process and give an account of it.

Process _____ **1**

Account **2**

_____

_____

_____

_____

3. The diagram below shows the transfer of chemical energy between metabolic pathways.

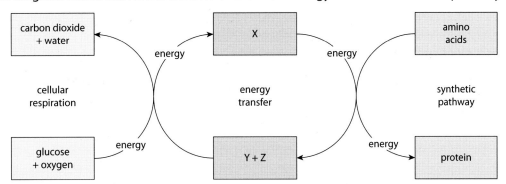

(a) Name substances X, Y and Z. **2**

X _____

Y _____

Z _____

(b) Name the type of metabolic pathway that releases energy, as represented by cellular respiration in this diagram. **1**

_____

(c) Describe the role of oxygen after ATP has been synthesised. **1**

_____

(d) The amino acids in the synthetic pathway are joined to form a chain.

Name the bond which joins amino acids to one another. **1**

_____

4. An experiment was carried out to investigate the effect of different respiratory substrates on the rate of respiration in yeast. To measure the rate of respiration, an indicator called methylene blue can be used and when it accepts hydrogen ions, it changes from dark blue to colourless.

A student set up three test tubes as shown below. Each tube contains 10 cm³ yeast, 10 cm³ methylene blue and 10 cm³ of the substrate indicated.

The experiment was left for one hour.

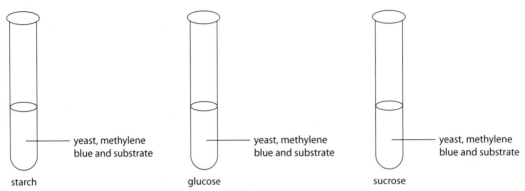

yeast, methylene
blue and substrate

starch

yeast, methylene
blue and substrate

glucose

yeast, methylene
blue and substrate

sucrose

The results from each test tube are in the table below.

Respiratory substrate	Colour of methylene blue after one hour
starch	dark blue
glucose	colourless
sucrose	light blue

(a) Name **two** variables, not already mentioned, that should be kept constant.   1

_____

_____

(b) State **one** way in which the reliability of this investigation could be improved.   1

_____

_____

(c) Explain why the experiment was left for one hour before the results were recorded.   1

_____

_____

4. (continued)

(d) The student also added a test tube with pure water instead of a substrate, all other contents identical.

State the term used to describe this experiment and explain its purpose.

Term _____ **1**

Purpose **1**

_____

_____

(e) A similar experiment was carried out with different glucose concentrations and the time taken to change to colourless for each concentration noted. The results are shown below.

Glucose concentration (%)	Time taken for methylene blue to change to colourless (seconds)
1	90
2	64
3	59
4	52
5	46
6	38
7	30

Plot a line graph to illustrate the results of the experiment. **2**

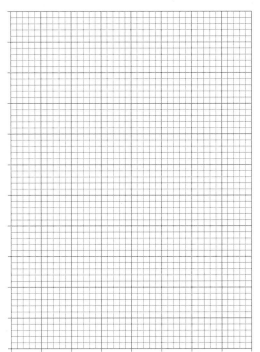

(f) State **one** conclusion that can be drawn from the results of this experiment.   **1**

_____

_____

5. The diagram below shows the structure of an artery.

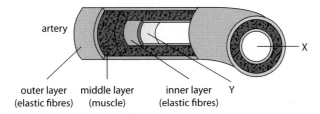

artery

outer layer (elastic fibres)   middle layer (muscle)   inner layer (elastic fibres)   Y   X

(a) State the names of X and Y.   **2**

X _____

Y _____

(b) Give **one** function of the elastic walls of the arteries.   **1**

_____

_____

(c) Changes in the smooth muscle can control blood flow through an artery.

State the term used to describe one of these changes and give an explanation of how the change affects blood flow in the vessels supplying the capillaries.

Term _____   **1**

Explanation   **1**

_____

_____

(d) Give an account of the structural and functional differences between an artery and vein. **3**

_____

_____

_____

_____

_____

_____

6. The diagram below shows a sphygmomanometer which measures blood pressure.

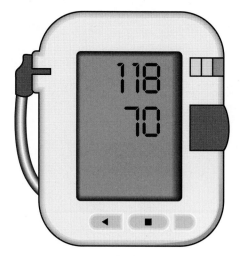

(a) State why there are two values shown on the meter. **1**

_____

_____

(b) State the typical values for a young adult. **1**

_____

(c) High blood pressure over a prolonged period can result in hypertension.

State **one** disease in which hypertension is a risk factor. **1**

_____

7. An experiment was carried out on 20 volunteers, who were split into two groups of ten.

Group 1 were asked to count to 26.

Group 2 were asked to recite the alphabet.

Both groups were then asked to read aloud what they saw on the card below.

The results are shown in the table below.

Group	Number of people giving each response	
	B	13
1	2	8
2	9	1

(a) Describe how the procedures used in this experiment led to the results shown. **2**

_____

_____

(b) Describe why the results of this experiment may not be considered reliable. **1**

_____

_____

(c) State what is meant by perceptual set. **1**

_____

_____

(d) State the **three** factors which may affect perceptual set. **2**

1 _____

2 _____

3 _____

8. The diagram below shows the inheritance of a sex-linked condition in a family.

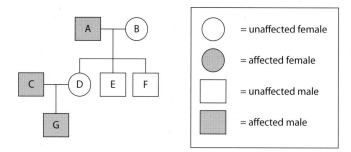

= unaffected female

= affected female

= unaffected male

= affected male

(a) The condition is caused by a recessive sex-linked allele represented by the letter g.

State the genotypes of individuals C and D. **1**

Individual C _____

Individual D _____

(b) Explain why individual A could not pass the condition to his sons. **1**

_____

_____

(c) Individual F has a son with a woman who is a carrier of the condition.

Calculate the percentage chance of their son having this condition. **1**

Space for calculation

_____%

(d) The condition is caused by a change in the genome in which an extra nucleotide is inserted into the gene that codes for an enzyme.

State the general term that is used to describe a gene alteration. **1**

_____

(e) Explain the likely effect of this extra nucleotide on the structure of the enzyme. **2**

_____

_____

_____

_____

(f) The condition occurs with a frequency of 1 in 700 males.

Assuming an equal proportion of males and females, calculate how many males are likely to have the condition in a town with a population of 350,000. **1**

Space for calculation

_____

(g) Where conditions such as this exist in a family, the family history can be used to determine the genotypes of its individual members.

Give the term used for this process. **1**

_____

(h) During *In vitro* fertilisation (IVF) treatment, it is possible to detect single gene disorders in fertilised eggs before they are implanted into the mother.

State the term that describes this procedure. **1**

_____

9. The table below shows the number of people with type 1 or 2 diabetes in four countries of the world in 2015 and the predicted number of people with diabetes in 2040.

Country	Total population in millions (2015)	Number of people with diabetes in millions (2015)	Projected number of people with diabetes in millions (2040)
China	1,375	110	150
India	1,311	70	124
USA	320	30	35
Brazil	208	14	23

(a) Calculate the country that had the highest percentage of people with diabetes in 2015.   **1**

Space for calculation

Country _____

(b) Calculate the simple whole number ratio for the number of people who have diabetes in the USA in 2015 and the projected number of people with diabetes in the USA in 2040.   **1**

Space for calculation

_____ : _____

(c) Tick the correct box(es) in the table below to show which statements refer to type 1 diabetes, type 2 diabetes or both.   **2**

Statement	Type 1 diabetes	Type 2 diabetes
Usually occurs in childhood		
Typically develops later in life		
Occurs mainly in overweight individuals		
Individual unable to produce insulin		
Can be treated with regular doses of insulin		
Individuals produce insulin but their cells are less sensitive to it		
Blood glucose levels rise rapidly after a meal		

10. A student carried out an investigation to determine the memory span for numbers.

Six children from four different age groups were each given 30 minutes to memorise a sequence of numbers and then recall as many as they could, in the correct order.

The results that they achieved are shown in the table below.

Child	Numbers recalled			
	Age 6–8	Age 9–11	Age 12–14	Age 15–17
1	7	9	17	27
2	9	14	15	13
3	3	12	24	9
4	6	10	8	14
5	12	19	16	10
6	5	2	22	23
Average	7		17	16

(a) Calculate the average numbers recalled by the children aged 9–11 years and write your answer **in the table** above. **1**

Space for calculation

(b) State **one** conclusion that can be drawn from the results of this investigation. **1**

_____

_____

(c) Describe **one** variable that would have to be kept constant to ensure a valid comparison could be made between the four groups of children. **1**

_____

_____

(d) Give **one** possible reason why children in the age group 12–14 years had a higher average than children in the age group 15–17 years. **1**

_____

_____

10. (continued)

(e) Describe **one** way in which the student could adapt her investigation to ensure that these results are reliable. **1**

_____

_____

(f) The children were then asked to repeat the experiment in front of an audience. This caused an improvement in their number recall.

State the term used to describe this improvement in performance. **1**

_____

(g) Some of the children with the higher scores were then rewarded and they agreed to take part in the experiment again. The children who were not rewarded did not wish to take part again.

State the term used to describe this. **1**

_____

11. The table below shows the volume of blood, measured in cm^3, in a man's right ventricle at different times during one cardiac cycle.

Time (s)	Volume of blood (cm³)
0·0	125
0·1	148
0·2	103
0·3	70
0·4	56
0·5	55
0·6	98
0·7	125

(a) Use the data **in the table** to calculate the man's heart rate in beats per minute. **1**

Space for calculation

_____ beats per minute

(b) If a second person has the same stroke volume and a heart rate of 54, calculate their cardiac output in cm^3/min. **1**

Space for calculation

_____ cm³/min

(c) Use the information above to complete the table below to show whether the valves are open or closed at each of the times shown.

Complete the table below by writing open or closed in the appropriate boxes. **2**

Time (s)	Valve between right atrium and right ventricle	Valve between right ventricle and pulmonary artery
0·2		
0·6		

12. Answer either A or B in the space below.

Labelled diagrams may be used where appropriate.

A   Describe hormonal control under the following headings:

    (i)   onset of puberty;   3

    (ii) control of sperm production;   3

    (iii) control of the menstrual cycle.   4

OR

B   Describe infectious diseases under the following headings:

    (i)   cause and transmission;   5

    (ii) spread and control.   5

Paper B

Higher Human Biology

Paper B

You are allowed 2 hours and 30 minutes to complete this paper.

A total of 100 marks are available in this paper.

Section 1 – 20 marks

Section 2 – 80 marks

Attempt ALL questions.

Leckie ✕ Leckie

Scotland's leading educational publishers

SECTION 1 – 20 marks

Attempt ALL questions.

Answers should be given on the separate answer sheet provided.

1. Which of the following shows the correct structure of RNA?

	Bases	Strand(s)	Sugar
A	ATGC	single	deoxyribose
B	AUGC	double	deoxyribose
C	ATGC	double	ribose
D	AUGC	single	ribose

2. Proteins are formed by chains of amino acids.

   Name the type of bond that joins amino acids into a chain.

   A Peptide

   B Hydrogen

   C Sugar–phosphate

   D Disulphide.

3. In one experiment, amplification of DNA by PCR commenced with 500 DNA molecules in the reaction tube.

   How many DNA molecules were present after five cycles of PCR?

   A 1,600

   B 2,500

   C 8,000

   D 16,000

4. Which of the following does **not** describe catabolism?

   A Releases energy

   B Involves the breakdown of molecules

   C Has reversible and irreversible steps and alternative routes

   D Involves biosynthetic processes.

5. The diagram below represents a mitochondrion which has been magnified 100 times.

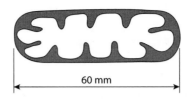

60 mm

What is the actual length of this mitochondrion? (1 mm = 1000 micrometres)

A   0·6 micrometres

B   6 micrometres

C   60 micrometres

D   600 micrometres.

6. The diagram below shows the citric acid cycle.

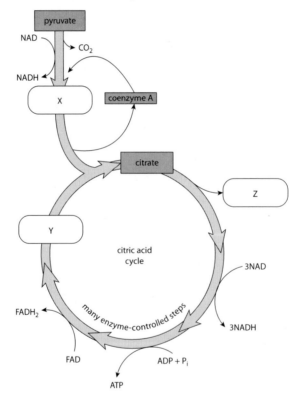

Which of the following correctly identifies X, Y and Z?

	X	Y	Z
A	acetyl coenzyme A	oxaloacetate	hydrogen
B	glucose	acetyl coenzyme A	oxaloacetate
C	acetyl coenzyme A	oxaloacetate	carbon dioxide
D	glucose	carbon dioxide	oxaloacetate

7. Which of the following correctly matches the reproductive organ, tissue or cell with its correct function in the reproductive tract?

	Organ/Tissue/Cell	Function
A	seminiferous tubules	produce testosterone
B	prostate gland	produce sperm
C	interstitial cells	secrete luteinising hormone (LH)
D	seminal vesicles	secrete fluids that maintain the mobility and viability of the sperm

8. Which of the following treatments for infertility involves the head of the sperm being drawn into a needle and injected directly into the egg to achieve fertilisation?

A   Artificial insemination

B   Intra-cytoplasmic sperm injection (ICSI)

C   *In vitro* fertilisation (IVF)

D   Stimulating ovulation.

9. Which of the following antenatal tests can produce cells for culturing to produce a karyotype for the diagnosis of a range of conditions?

A   Amniocentesis

B   Rhesus antibody testing

C   Biochemical tests

D   Ultrasound imaging.

10. The diagram below shows the heart during three phases of the cardiac cycle.

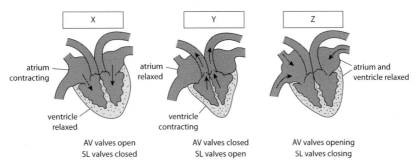

Which of the following correctly identifies phases X, Y and Z?

	X	Y	Z
A	atrial systole	diastole	ventricular systole
B	ventricular systole	atrial systole	diastole
C	atrial systole	ventricular systole	diastole
D	diastole	ventricular systole	atrial systole

11. Which of the following terms describes a clot which breaks loose and travels through the bloodstream until it blocks a blood vessel?

   A   Thrombus

   B   Fibrin

   C   Embolus

   D   Atheroma.

12. Scientists conducted an investigation into the effect of drinking different amounts of alcohol on the risk of developing heart disease.

The graph shows the results of the investigation.

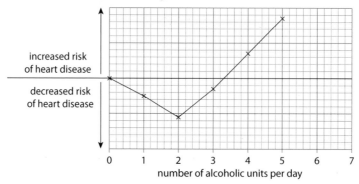

How many alcoholic drinks per week would result in the greatest chance of heart disease?

   A   2

   B   5

   C   6

   D   35

13. Doctors investigated the effect of taking glucose by mouth compared with injecting the same mass of glucose directly into the blood of type 2 diabetic people and non-diabetic people.

The doctors:

- investigated type 2 diabetic people and non-diabetic people

- monitored the concentration of insulin in the blood for three hours following the intake of glucose.

The doctors used both type 2 diabetic people and non-diabetic people in this study. This

   A   allowed a valid comparison

   B   increased reliability

   C   increased accuracy

   D   ensured the variables were kept constant.

14. Name the three elements which have an effect on perceptual set and the way a stimulus is perceived.

A   Past experience, context and expectation

B   Past experience, context and recognition

C   Perception of distance, segregation of objects and expectation

D   Perception of distance, segregation of objects and recognition.

15. The diagram below shows a motor neuron.

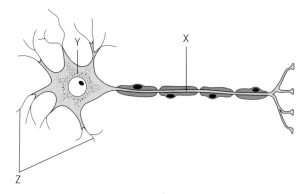

Which of the following correctly identifies structures X, Y and Z?

	X	Y	Z
A	cell body	dendrite	axon
B	axon	dendrite	cell body
C	dendrite	axon	cell body
D	axon	cell body	dendrite

16. Which of the following descriptions does **not** apply to authoritative control?

A   Sets limits, rules, high standards and explains the reasons for them

B   Explains the consequences of unacceptable behaviour

C   Allows a child to regulate their own behaviour

D   Gives direction and expects responsible behaviour and cooperation in return.

17. The diagram below shows an immune response.

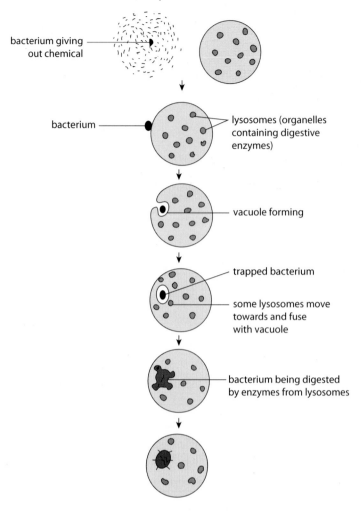

Name the type of cell which carries out this immune response.

A   Phagocyte

B   Natural killer cell

C   T lymphocyte

D   B lymphocyte.

18. Which of the following design features would be used to eliminate bias from clinical trials?

A   Use of placebo

B   Randomisation

C   Large sample size

D   Replicates.

19. The table below contains information about the number of cases of tuberculosis (TB) in a city over five years.

Year	TB cases in June	TB cases in December
2009	220	780
2010	445	620
2011	450	1200
2012	135	125
2013	150	400

Which of the following conclusions can be drawn from the data in the table?

A   There are always more cases of TB in December than in June

B   The number of cases of TB increased by 60% between June and December of 2010

C   The greatest percentage decrease in TB cases occurred between June of 2010 and December of 2010

D   The greatest percentage increase in TB cases occurred between June 2009 and December 2009.

20. Whooping cough is a disease that affects some infants. Doctors collected data relating to whooping cough between 1965 and 1995.

They collected data for:

- the number of cases of whooping cough reported

- the percentage of infants vaccinated against whooping cough.

The graph shows the data collected by the doctors.

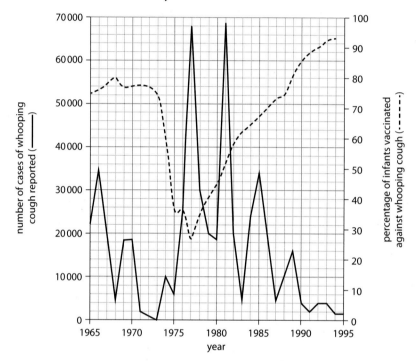

Which of the following best describes the general relationship between the number of cases of whooping cough reported and the percentage of infants vaccinated against whooping cough?

When the percentage of infants vaccinated is above

A   50% the number of cases is always below 40,000

B   50% the number of cases is always above 40,000

C   60% the number of cases is always above 40,000

D   60% the number of cases is always below 40,000.

SECTION 2 – 80 marks

Attempt ALL questions.

It should be noted that question 12 contains a choice.

1. The diagram below shows two strands of DNA.

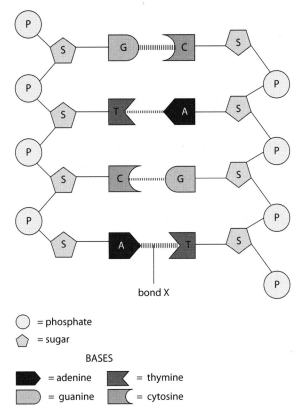

bond X

○ = phosphate

⬠ = sugar

BASES

◗ = adenine   ◀ = thymine

▭ = guanine   ◖ = cytosine

(a) Name bond X.   **1**

_____

(b) Give the term used to describe the unit made up of sugar, phosphate and base.   **1**

_____

(c) Name the sugar found in a strand of DNA.   **1**

_____

(d) Describe the directional arrangement of the two strands in DNA.   **2**

_____

_____

1. (continued)

(e) During DNA replication, bond X is broken.

Name the enzyme which does this. **1**

_____

(f) Name **two** molecules, not shown on the diagram or mentioned in your answer above, which are essential for DNA replication. **2**

_____

_____

(g) Explain why DNA replication must take place before a cell divides. **1**

_____

_____

2. Below is a sequence of bases from a volunteer's DNA who had her entire genome sequenced.

TAGTTGACCATGCATGTTCAGAGC

(a) State the term used for the use of computer technology to identify the sequence of bases shown. **1**

_____

(b) Systematics compares human genome sequence data and genomes of other species.

Explain why this is useful. **1**

_____

_____

(c) Personalised medicine is based on an individual's genome.

Give **one** advantage that personalised medicine may have. **1**

_____

_____

(d) DNA probes are used to detect the presence of specific sequences in samples of DNA, such as the one above.

Describe the structure of a DNA probe. **1**

_____

_____

3. During strenuous muscular activity, the cell rapidly breaks down its reserves of ATP to release energy. Muscle cells have an additional source of energy which can be used to replenish ATP pools during strenuous bouts of exercise.

(a) State the name of the substance found in muscle cells which can act as an additional source of energy. **1**

_____

(b) During strenuous exercise, the muscle cells do not get sufficient oxygen to support the electron transport chain. Under these conditions, describe the conversion that occurs within the muscle cells. **1**

_____

_____

(c) Explain what is meant by the term 'oxygen debt'. **2**

_____

_____

(d) Samples of muscle cells from two different athletes were taken and are shown below.

A   B

State which of the two cells belongs to an endurance athlete and give a reason for your selection.

Cell _____ **1**

Reason _____ **1**

_____

3. (continued)

(e) There are two types of muscle fibres.

Name both and describe **two** differences between them.

Types of muscle fibres _____   **1**

_____

Differences 1. _____   **2**

_____

2. _____

_____

4. Obesity is a major risk factor for cardiovascular disease and type 2 diabetes. The table below shows the incidence rate of type 2 diabeties (per 100,000 population per year) in Scotland between 2007 and 2014 in different age ranges (in years).

Incidence rate of type 2 diabetes by age.

Age range	2007	2008	2009	2010	2011	2012	2013	2014
<10	0	0	0	0	0	0	0	0
10–19	4	3	4	3	2	3	1	3
20–29	23	24	24	24	23	29	25	27
30–39	101	127	120	116	113	146	124	115
40–49	304	310	335	321	322	347	323	315
50–59	608	628	653	629	595	650	606	570
60–69	955	917	944	851	814	833	783	733
≥ 70	835	874	867	799	747	729	762	649
Average	339	350	360	338	325	343	336	307

(a) Describe the relationship between age and incidence of type 2 diabetes in 2014.　　**2**

_____

_____

_____

(b) Give one possible reason for the relationship between age and incidence of type 2 diabetes, as described in (a).　　**1**

_____

_____

(c) Calculate the percentage change in type 2 diabetes incidence for those aged 30–39 between 2007 and 2010.　　**1**

Space for calculation

_____%

(d) Predict the incidence of type 2 diabetes for those aged 60–69 in 2015.　　**1**

Space for calculation

_____ per 100,000 population per year.

5. Many couples can conceive a child within a few menstrual cycles but some couples are unable to. This could be due to a fertility problem with either the male or female reproductive systems.

In vitro fertilisation (IVF) is a widely used treatment for infertility.

At one IVF clinic, over 1,000 treatment cycles were monitored. The percentage of live births per treatment cycle was recorded.

The results were recorded against the age of the women and are shown below.

Age of women (years)	Percentage of live births per treatment cycle
Under 34	27·6
34–36	22·3
37–39	18·3
40–42	10·0
Over 42	5·0

(a) Describe the general trend shown in this table. **1**

_____

_____

(b) Calculate the percentage decrease in percentage of live births per treatment cycle between women under 34 years compared with 40 to 42 years. **1**

Space for calculation

_____%

(c) Describe one reason that couples may experience difficulties conceiving due to infertility. **1**

_____

_____

(d) Aside from IVF, describe **one** other fertility treatment that couples may use. **1**

_____

(e) Couples who do not wish to conceive a child may use contraception.

State **one** example of contraception. **1**

_____

6. Each year, between 1980 and 2000, health workers in an African country recorded:

- the percentage of the population infected with the human immunodeficiency virus (HIV)

- the number of new cases of tuberculosis (TB) per 100,000 people per year.

The results are shown below.

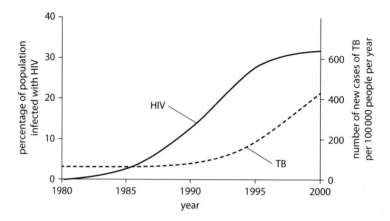

(a) The number of new cases of tuberculosis (TB) per year was recorded as number per 100,000 of the population.

Give **one** reason why. **1**

_____

_____

(b) Give an account of how HIV and TB affect the immune system. **3**

_____

_____

_____

_____

7. Ten students took part in an experiment where they were asked to complete ten Sudoku puzzles. They completed the puzzles in a room alone.

Five of the students were rewarded after they completed each individual puzzle with a verbal announcement of 'well done' from a speaker. The other five received no such reward.

The time taken in seconds for each student to complete each Sudoku puzzle was recorded.

The results of the experiment are shown below.

Rewarded

Student	Time to complete Sudoku puzzle (seconds)									
	1	2	3	4	5	6	7	8	9	10
1	305	270	280	253	207	202	185	190	187	190
2	506	498	456	434	423	410	407	402	410	415
3	435	406	340	395	376	362	290	307	298	250
4	620	630	606	580	567	550	520	404	397	456
5	107	140	132	116	114	127	143	103	097	109
Average	395	389	363	356	338	331	310	283	280	284

Not rewarded

Student	Time to complete Sudoku puzzle (seconds)									
	1	2	3	4	5	6	7	8	9	10
6	256	243	232	210	214	230	237	238	240	247
7	403	385	356	345	354	340	370	362	389	392
8	506	482	493	453	467	432	408	427	432	456
9	197	180	156	145	134	130	156	145	125	121
10	150	145	134	187	123	134	120	145	132	147
Average	303	287	275	269	259	254	260	263	265	275

(a) Calculate which group, rewarded or not rewarded, showed the greatest percentage decrease in the time taken to complete the Sudoku puzzle. **1**

Space for calculation

Group _____

(b) State the student with the lowest individual time across the ten attempts. **1**

_____

7. **(continued)**

(c) Describe the overall trend for both groups. **1**

_____

_____

(d) Name **one** variable that should be controlled when setting up this experiment. **1**

_____

(e) Name **one** variable that should be kept constant during this experiment. **1**

_____

(f) State **one** way in which the reliability of the results from this investigation could be improved. **1**

_____

_____

(g) Draw a bar graph to illustrate the average results of the rewarded experiment. **2**

280

7. (continued)

(h) State **one** conclusion that can be drawn from the results of this experiment. **1**

_____

_____

8. Scientists investigated the effect of a 6-week exercise programme on the resting heart rate of males and females.

The scientists recruited a large group of male volunteers and a large group of female volunteers. They measured the resting heart rate, in beats per minute, of each volunteer before the exercise programme. Both groups took part in the same exercise programme.

The scientists measured the resting heart rate of each volunteer after the exercise programme.

The scientists determined the mean resting heart rate and the range of resting heart rates for each group before and after the exercise programme.

The graph below shows their results.

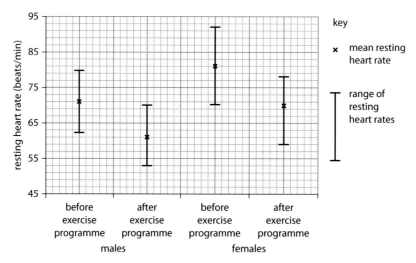

(a) Give the range of the resting heart rates in males after the exercise programme in beats per minute. **1**

_____ beats per minute

(b) Calculate the percentage decrease in the mean resting heart rate of females after the exercise programme. **1**

Space for calculation

_____%

(c) The scientists used the percentage change in the mean resting heart rate after the exercise programme to compare the results for males and females.

Explain why they used percentage change in the mean resting heart rate. **1**

_____

_____

8. (continued)

(d) The scientists calculated the cardiac output of the volunteers before and after the exercise programme. In some volunteers, their cardiac output stayed the same, even though their resting heart rate decreased.

Explain how their cardiac output could stay the same even when their resting heart rate had decreased.　　**1**

_____

_____

9. 40 students took part in an investigation to test their short-term memory.

They were asked to listen to a digital pre-recording of a series of random letters (i.e. that did not spell a word). They then had to write down the letters that they heard, but only once the sequence was complete. Their scores were grouped together and recorded below.

Number of letters in sequence read out	Number of students who recalled all the letters in the correct order
1	40
2	40
3	39
4	40
5	32
6	27
7	19
8	7
9	5
10	3

(a) Draw a bar graph to illustrate the results of the experiment.   **2**

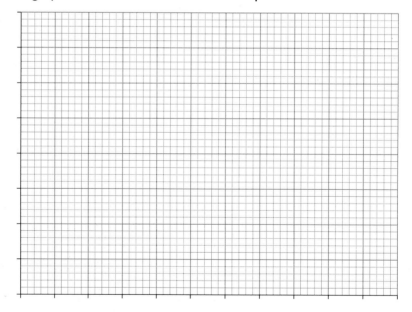

9. (continued)

(b) Predict the number of students who would correctly recall a sequence with 11 letters in it.   **1**

_____

(c) Using data **from the graph**, describe the relationship between the number of letters in the sequence and the number of students who recalled all the letters in the correct order.   **1**

_____

_____

(d) Calculate, as a simple whole number ratio, the number of students who recalled all the letters in the correct order when four letters were in the sequence compared to when five letters were in the sequence.   **1**

Space for calculation

_____:_____

(e) Explain why a pre-recording was used rather than someone reading aloud live.   **1**

_____

_____

(f) Explain why 40 students were used and their results grouped together.   **1**

_____

_____

(g) State **one** method which could be used to improve short-term memory.   **1**

_____

(h) State the term used to describe the loss of items from short-term memory.   **1**

_____

10. 1,000 volunteers took part in a clinical trial for Alzheimer's disease.

500 volunteers were given a trial medication for six months while the other 500 were given a tablet which did not contain the trial medication.

A computer program allocated the volunteers to one group or another and neither the volunteers nor the doctors knew which group they were in.

(a) State the name given to the tablet which did not contain the trial medication. **1**

_____

(b) Explain the purpose of giving this tablet to half the group. **1**

_____

_____

(c) State the term used to describe a trial where neither the doctors nor the volunteers know which group they are allocated to. **1**

_____

(d) Explain why a computer program, rather than the doctors, was used to allocate the volunteers. **1**

_____

_____

(e) Explain why a large group size was chosen for this trial. **1**

_____

_____

11. A range of white blood cells constantly circulates, monitoring the tissues. If tissues become damaged or invaded, cells release cytokines which increase blood flow, resulting in specific white blood cells accumulating at the site of infection or tissue damage.

(a) Name the process of constant monitoring by white blood cells.   **1**

_____

(b) Describe the main events that occur in clonal selection.   **2**

_____

_____

_____

(c) Type 1 diabetes is an auto-immune disease in which T-cells attack insulin producing cells in the pancreas.

Describe why this happens.   **2**

_____

_____

(d) Many people suffer from allergies such as hay fever.

Explain what happens in an allergic response.   **2**

_____

_____

_____

12. Answer either A or B in the space below.

Labelled diagrams may be used where appropriate.

A   Make notes under the following headings:

    (i)   neurotransmitters at synapses;   **5**

    (ii) neurotransmitters, mood and behaviour.   **5**

OR

B   Describe heart structure and function under the following headings:

    (i)   the cardiac cycle;   **5**

    (ii) cardiac conducting system.   **5**

Paper C

Higher Human Biology

Paper C

You are allowed 2 hours and 30 minutes to complete this paper.

A total of 100 marks are available in this paper.

Section 1 – 20 marks

Section 2 – 80 marks

Attempt ALL questions.

Scotland's leading educational publishers

SECTION 1 – 20 marks

Attempt ALL questions.

Answers should be given on the separate answer sheet provided.

1. If 20% of the bases in a DNA molecule are cytosine, what is the ratio of thymine to cytosine in this molecule?

   A   2:3

   B   3:2

   C   1:1

   D   2:1

2. A mature transcript is made up of

   A   RNA composed of exons and introns

   B   DNA composed of exons and introns

   C   RNA composed of exons only

   D   RNA composed of introns only.

3. A nonsense gene mutation

   A   changes one codon into another but does not affect the protein produced

   B   stops translation completely so no protein is produced

   C   changes a codon into a stop codon and shortens the protein produced

   D   is a substitution and the protein produced functions in a different way.

4. The results of a paternity test are shown below and show DNA from six people.

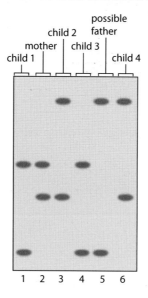

Which of the following conclusions is correct?

A   Children 1 and 3 cannot be the biological offspring of this father

B   Children 2 and 3 cannot be the biological offspring of this father

C   None of the children can be the biological offspring of this father

D   All of the children could be the biological offspring of this father.

5. The diagram below shows the action of a competitive inhibitor in an enzyme-catalysed reaction.

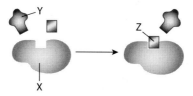

Which of the following correctly identifies the molecules involved in this reaction?

	Enzyme	Inhibitor	Substrate
A	X	Y	Z
B	Y	Z	X
C	Z	X	Y
D	X	Z	Y

6. The following diagram shows the sequence of amino acids in a small protein molecule.

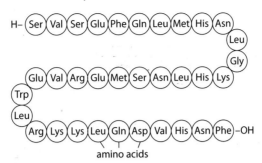

The ratio of histidine (His) : leucine (Leu) : glycine (Gly) : valine (Val) is

A   3 : 2 : 3 : 3

B   3 : 5 : 3 : 3

C   3 : 2 : 3 : 1

D   2 : 2 : 3 : 3

7. A technician needs to make up 200 cm³ of 0·9% salt solution from a 1·8% stock solution.

She can do this by

A   adding 100 cm³ of water to 100 cm³ of the stock solution

B   adding 50 cm³ of water to 150 cm³ of the stock solution

C   diluting the stock solution ten times

D   concentrating the stock solution ten times.

8. The graph below shows the trace of a human heart beat, known as an electrocardiogram (ECG).

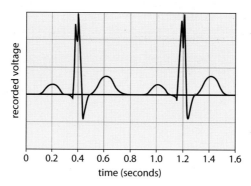

What is the heart rate in beats per minute?

A   65

B   70

C   75

D   80

9. Which of the following correctly shows the function of some parts of the male reproductive system?

	Seminiferous tubules	Interstitial cells	Prostate gland
A	sperm formation	secretion of testosterone	secretion of thin liquid which forms part of semen
B	sperm formation	secretion of thin liquid which forms part of semen	secretion of testosterone
C	secretion of thin liquid which forms part of semen	secretion of testosterone	sperm formation
D	secretion of testosterone	sperm formation	secretion of thin liquid which forms part of semen

10. The following events occur during the menstrual cycle.

1　Follicle develops into a corpus luteum

2　Follicle-stimulating hormone stimulates follicle to mature

3　Corpus luteum secretes progesterone

4　Oestrogen from follicle stimulates repair of the endometrium.

The sequence in which these events occur is

A　4, 2, 1, 3

B　1, 2, 3, 4

C　4, 2, 3, 1

D　4, 3, 2, 1

11. Which of the following descriptions is correct for arteries and veins?

	Arteries	Veins
A	relatively narrow lumen	thin layer of smooth muscle
B	thick layer of smooth muscle	no valves present
C	few elastic fibres	valves present
D	relatively wide lumen	thick layer of smooth muscle

12. Which of the following statements about high density lipoproteins (HDLs) and low density lipoproteins (LDLs) is correct?

A　High levels of HDLs are associated with high levels of blood cholesterol

B　Low levels of LDLs are associated with high levels of blood cholesterol

C　A high HDL : LDL ratio in the blood increases the chances of atherosclerosis developing

D　A low HDL : LDL ratio in the blood is linked to low levels of blood cholesterol.

13. Two students, John and James, were given some fruit juice to drink and the change in their respective blood glucose levels was checked every 30 minutes. The results are shown in the table below.

Time after drinking juice (minutes)	Blood glucose level (mg/100 cm³)	
	John	James
0	85	87
30	140	165
60	90	100
90	85	90

Which of the following statements is correct?

A   After 30 minutes, John's blood glucose level was twice his original

B   James' blood glucose level dropped by approximately 40% between 30 and 60 minutes after drinking the juice

C   After 60 minutes, James' blood glucose concentration was approximately 20% greater than John's

D   John took longer than James to return to his original blood glucose level.

14. A high blood glucose level

A   is detected by pancreatic receptors causing a decrease in insulin secretion

B   stimulates insulin secretion and the conversion of glucose into glycogen

C   is detected by pancreatic receptors causing an increase in glucagon secretion

D   causes a decrease in adrenaline production and an increase in insulin production.

15. The limbic system

A   regulates fear, anger, aggression and does not involve the autonomic nervous system

B   may be involved in some forms of addiction and does not involve the endocrine system

C   is important for the formation and transfer of new memories

D   is not linked to hunger or thirst responses.

16. The graph below shows the effect of repetition (number of practice trials) on the time taken to perform a particular task.

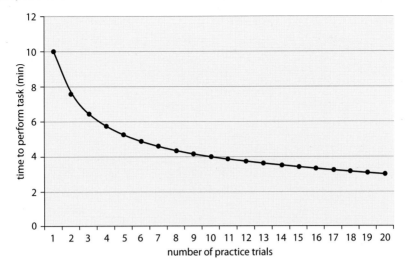

Which of the following is a valid conclusion from this data?

A   The improvement in performance was uniform over the trial period

B   Practising this particular task does not improve performance

C   Improvement was most evident in the early trials

D   Continued trials would reduce the time taken to almost zero.

17. A boy was bitten by a pet snake. Now he is afraid of all snakes. This is a demonstration of

A   deindividuation

B   social facilitation

C   generalisation

D   discrimination.

18. Apoptosis

A   does not involve the synthesis of self-destructive enzymes

B   is a form of programmed cell death

C   causes cells to release harmful substances into the surrounding area

D   is a specific defence mechanism.

Questions 19 and 20 refer to the following graph.

The graph shows the effect on body temperature, measured in °C, of immersing the arm of a volunteer in warm water for five minutes.

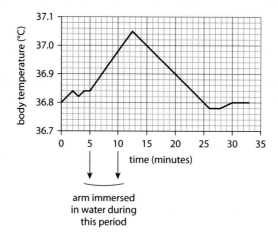

19. What was the range of the volunteer's body temperature from five minutes after the start of the investigation until 30 minutes after the start of the investigation?

   A   2·7°C

   B   2·5°C

   C   0·27°C

   D   0·25°C

20. Predict the volunteer's body temperature after 35 minutes if no further procedure was carried out.

   A   36·76°C

   B   36·78°C

   C   36·82°C

   D   36·80°C

SECTION 2 – 80 marks

Attempt ALL questions.

It should be noted that question 14 contains a choice.

1. The diagram below shows an abnormal pattern of cell division causing cancer.

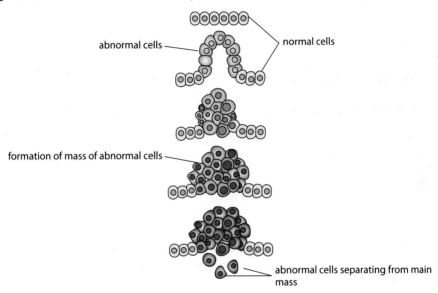

(a) State **one** reason why such a mass of abnormal cells might form. **1**

_____

_____

(b) State the term used for this mass of abnormal cells. **1**

_____

(c) Give **one** possible reason why some abnormal cells may fail to attach to each other and separate from the main mass. **1**

_____

_____

(d) State **one** possible effect the separation of some abnormal cells from the main mass might have on the development of the cancer. **1**

_____

2. In humans, the amino acid phenylalanine can normally be converted by the enzyme phenylalanine hydroxylase (PAH) into the amino acid tyrosine. If there is an excess of phenylalanine, it is either converted to tyrosine by PAH in this way or excreted in the urine.

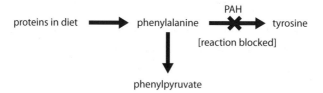

Some people cannot make PAH and so phenylalanine is changed by other enzymes into potentially harmful substances which include phenylpyruvate.

If phenylpyruvate accumulates, especially in very early life, the mental development of the child is at permanent risk. This condition is called phenylketonuria.

(a) (i) Give **one** way in which the genetic code for the synthesis of phenylalanine hydroxylase in someone affected by phenylketonuria could be changed. **1**

_____

_____

(ii) Explain why this changed genetic code no longer allows for the synthesis of phenylalanine hydroxylase. **1**

_____

_____

(b) Phenylketonuria can be detected by a simple blood test very soon after birth. Suggest **one** benefit of this early detection. **1**

_____

_____

2. **(continued)**

(c) Suggest **one** strategy which could eliminate the impact of the lack of the enzyme phenylalanine hydroxylase. **1**

_____

_____

(d) Phenylketonuria is both an autosomal and a recessive trait.

Explain each of these two features of the trait in the inheritance pattern which would be shown in a family history involving this condition. **2**

Autosomal _____

_____

_____

Recessive _____

_____

_____

3. A student designed the following experiment to compare the activity of the enzyme catalase in four different tissues, A, B, C and D.

Equal masses of each tissue were mixed with equal volumes of water and then blended in a food processor. After filtering the extracts, the student took 2 cm³ of each extract and added this to 5 cm³ of hydrogen peroxide. The volume of oxygen given off within a set time was measured and converted into an 'activity' value. The higher the activity value, the more oxygen was released. This was repeated twice more and the results are shown below as the 'relative activity' for each tissue.

Tissue	Relative activity			
	1	2	3	Average
A	1	0·9	0·8	0·9
B	3·7	3·9	3·5	3·7
C	6·6	6·8	3·7	
D	0·4	0·3	0·2	0·3

(a) Complete the table by calculating the average activity for tissue C.   **1**

   Space for calculation

(b) Draw a bar chart to show the average activity for each tissue.   **2**

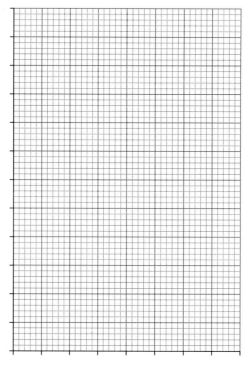

3. (continued)

(c) (i) State how the reliability of this experiment could be improved. **1**

_____

_____

(ii) State **two** variables which would need to be controlled to improve the validity of any conclusions. **2**

1 _____

2 _____

(iii) Suggest a suitable control for this experiment. **1**

_____

(iv) State a conclusion which can be drawn from these results. **1**

_____

_____

(v) Express as a whole number ratio the relative activity of tissue D compared to tissue A. **1**

Space for calculation

___ : ___

4. Estimating the number of healthy cells in a culture is important in determining the effect of certain drugs. Adenosine triphosphate (ATP) plays a vital role in energy exchange and its presence directly correlates with how efficiently cells are functioning.

A very successful 'bioluminescent' method for ATP determination uses the enzyme luciferase which catalyses the conversion of luciferin into oxyluciferin with the emission of light.

$$\text{ATP + luciferin} \xrightarrow{\text{luciferin}} \text{oxyluciferin + light}$$

(a) State **two** reasons why the emission of light is a good way of measuring ATP determination. **2**

_____

_____

_____

(b) State what a high light emission would indicate about how efficiently the cells being examined are functioning and explain your answer.

High light emission **1**

_____

_____

_____

Explanation **2**

_____

_____

_____

_____

(c) This bioluminescent method can detect very small quantities of ATP.

Give **one** feature this method must possess to enable it to detect such small quantities of ATP. **1**

_____

5. When skeletal muscle cells operate under conditions of oxygen deficiency, the following reaction takes place.

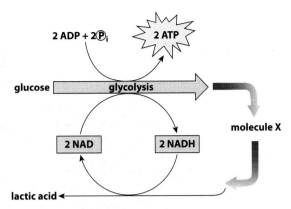

(a) Two molecules of ATP are formed by the addition of a phosphate group to each of two molecules of ADP.

State the term used to describe the addition of a phosphate group in this way.   **1**

_____

(b) Name molecule X.   **1**

_____

(c) During the conversion of molecule X to lactic acid, hydrogen is removed from NADH.

State the type of enzyme involved in this removal.   **1**

_____

6. When skeletal muscles operate, the fibres may contract quickly or slowly. These are called fast twitch fibres and slow twitch fibres respectively.

(a) Underline those activities which are best suited to slow twitch fibres.   **1**

   endurance sports – sprinting – pole vaulting – marathon running

(b) Name the type of fibre which has more mitochondria present.   **1**

_____

(c) State the type of fibre which is liable to fatigue quickly.   **1**

_____

(d) State the type of fibre which will be developed using a programme of regular weight-lifting.   **1**

_____

7. A scientist studied the different types of muscle fibres, fast twitch and slow twitch, in one thousand athletes in different types of sport. They also included in the study a sample from the average population of non-athletes.

The results are shown in the table below.

Sport	Number sampled	Average % fast twitch fibres	Average % slow twitch fibres
cycling	100	25	75
speed skating	250	30	70
canoeing	90	35	65
swimming	120	40	60
high jumping		75	25
average population	125	50	50

(a) Calculate how many high jumpers were sampled.   1

Space for calculation

(b) For any one sport, state **one** way in which the scientist could reduce the variation in the results obtained before the average was calculated.   1

_____

_____

(c) State the sport which had the highest ratio of fast twitch to slow twitch fibres.   1

_____

(d) State the sport which had a ratio of three slow twitch fibres to one fast twitch fibres.   1

_____

(e) State the sport which had three times as many fast twitch fibres as the group of cyclists.   1

_____

8. Down's trisomy is a genetic condition which causes various abnormal physical and mental changes during development.

The following diagram shows the chromosome complement of a Down's trisomy individual.

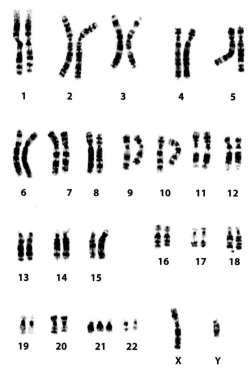

(a) State what is unusual about this chromosome complement.　　　　　　　**1**

_____

_____

(b) State the name given to a diagram, such as the one shown, which represents the chromosome complement of an individual.　　　　　　　**1**

_____

(c) From the diagram:

(i) State the sex of the individual.　　　　　　　**1**

_____

(ii) Describe the source of the cells used to produce such a diagram.　　　　　　　**1**

_____

_____

9. A woman who might be carrying a baby with a genetic condition can be offered a test to detect this during her pregnancy.

(a) State the general term used for such a test during pregnancy. **1**

_____

(b) One specific test is called amniocentesis.

Describe this test and the possible risk associated with it.

Description: _____ **3**

_____

_____

_____

Possible risk: _____ **1**

_____

10. The graph and table below show some of the results of a study carried out into the relationship between smoking and the second half of pregnancy.

The **graph** below shows the numbers and sex of babies born to mothers of different smoking habits.

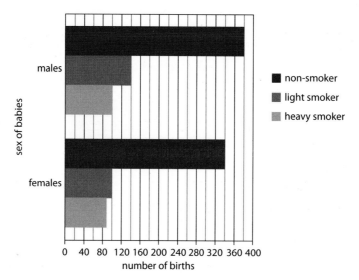

The **table** below shows some average measurements of the babies at birth born to mothers of different smoking habits.

Average measurement	Non-smoker	Light smoker	Heavy smoker
Birth mass (kg)	3·45	3·25	3·16
Circumference of head (cm)	36·00	35·80	35·60
Length (cm)	52·00	51·40	51·40

(a) Assuming a usual length of a pregnancy, state how long the study would last for each mother.   **1**

_____

(b) State how many babies were examined in total.   **1**

Space for calculation

_____

(c) State the percentage of mothers who were light smokers.   **1**

Space for calculation

_____%

10. (continued)

(d) State which of the average measurements is most affected when the mothers smoke heavily.　　　　**1**

_____

(e) State **two** trends in the average measurements shown by babies of mothers who do not smoke.　　　　**2**

1 _____

_____

2 _____

_____

(f) Suggest **two** uses which might be made of results from studies such as this.　　　　**2**

1 _____

_____

2 _____

_____

11. The diagram below shows a division of the human nervous system.

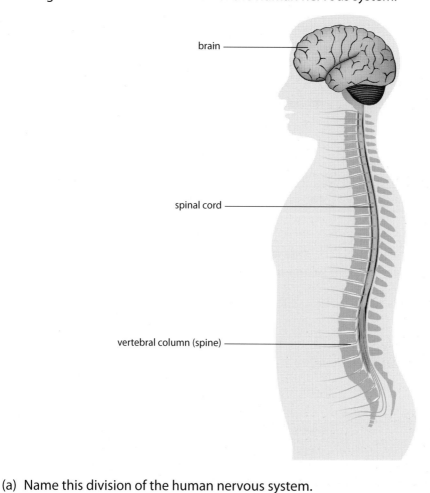

(a) Name this division of the human nervous system. **1**

_____

(b) Name the division of the human nervous system that controls heart rate. **1**

_____

(c) Name the type of cell, found in the spinal cord, which carries impulses towards muscles. **1**

_____

(d) Insert these terms into the correct boxes to show the direction that a nerve impulse travels in. The first one is done for you. **1**

cell body – effector – axon – dendrites – receptor

12. The person shown in the diagram below has had his corpus callosum surgically cut, separating his left and right cerebral hemispheres. He was asked to focus on a white spot in the middle of a screen then quickly shown the two images at the same time so that each image goes to the opposite hemisphere.

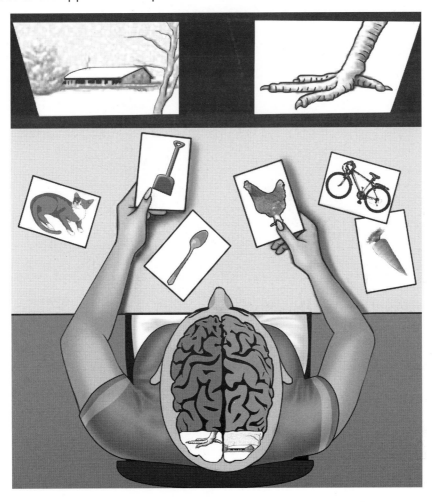

(a) Explain the function of the corpus callosum.   **1**

_____

_____

(b) (i)   The person was asked to say what he saw. Predict his answer.   **1**

_____

(ii) Now he was asked to lift up a card which corresponded to what he saw. Predict which card he would pick up.   **1**

_____

13. The diagram below shows a form of learning behaviour in human babies.

(a) State what this form of learning is called.   **1**

_____

(b) Give **one** example of this form of learning in adults.   **1**

_____

(c) Give **one** example of positive and **one** example of negative reinforcement of learning.

Positive reinforcement _____   **1**

_____

Negative reinforcement _____   **1**

_____

(d) Early infant attachment is important.

Give **one** reason for this attachment.   **1**

_____

_____

13. (continued)

(e) Explain what is meant by social facilitation. **1**

_____

_____

(f) Studies have shown that the use of video recording in football matches reduces the incidence of violent behaviour among rival fans.

In terms of deindividuation, give a possible explanation for this. **3**

_____

_____

_____

_____

_____

14. Answer either A or B in the space below.

Labelled diagrams may be used where appropriate.

A   Describe the male reproductive system under the following headings:

   (i)   structure and function of the testes;   **5**

   (ii) contribution of the prostate gland and seminal vesicles to fertilisation.   **5**

OR

B   Describe vaccination under the following headings:

   (i)   active immunity;   **5**

   (ii) herd immunity.   **5**

Paper D

Higher Human Biology

Paper D

You are allowed 2 hours and 30 minutes to complete this paper.

A total of 100 marks are available in this paper.

Section 1 – 20 marks

Section 2 – 80 marks

Attempt ALL questions.

Scotland's leading educational publishers

SECTION 1 – 20 marks

Attempt ALL questions.

Answers should be given on the separate answer sheet provided.

1. Which of the following descriptions is correct?

    A   Germline cells can only undergo meiotic cell division

    B   Somatic cells undergo mitosis to produce haploid cells

    C   Somatic cells undergo meiosis to produce diploid cells

    D   Stem cells are somatic cells which are relatively unspecialised.

2. The concentration of urea in a healthy person's urine was found to be 60 times as much as found in the blood.

    If a urine sample from such a person contained 6 g urea/litre, how much urea would be present in 100 cm^3 of blood?

    A   1·0 g

    B   0·1 g

    C   0·01 g

    D   0·001 g

3. The protein encoded by a gene may then have which of the following post-translational modifications?

    A   Cutting out amino acids from the polypeptide and the addition of a second polypeptide chain

    B   Removal of a phosphate or a carbohydrate group from a polypeptide chain

    C   Cutting mRNA and adding a carbohydrate group to the mRNA

    D   Removing either a phosphate or a carbohydrate group from the mRNA.

4. Which of the following shows the effect of a single gene mutation?

A	single nucleotide added to DNA	minor effect on protein synthesised	causes frameshift mutation
B	single nucleotide removed from DNA and replaced with a different nucleotide	causes frameshift mutation	has major effect on protein synthesised
C	single nucleotide added to DNA	has major effect on protein synthesised	does not cause frameshift mutation
D	single nucleotide removed from DNA and replaced with a different nucleotide	single nucleotide added to DNA	has minor effect on protein synthesised

5. An experiment was carried out which looked at how long it took for the same masses of five different drugs, used as painkillers, to dissolve in stomach acid. Conditions of temperature and pH were kept constant throughout.

The results are shown in the graph below.

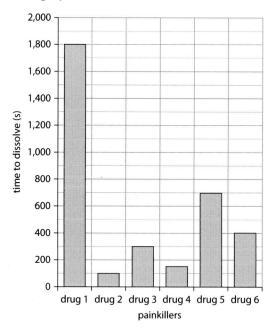

Which of the following conclusions would be valid from this data?

A   Drug 2 would be best at treating pain

B   Drug 6 dissolves faster than drug 5

C   Painkillers do not dissolve well in water

D   Drug 2 takes the least time to dissolve.

6. Which statement about enzyme inhibition is correct?

A In competitive inhibition, the inhibitor binds to the active site of the enzyme

B Non-competitive inhibitors are very similar in structure to the enzyme's normal substrate

C In competitive inhibition, a poison binds to the enzyme so that it can never work again

D All inhibitors act in an irreversible fashion.

7. The function of the high energy electrons in the electron transport chain is to

A combine with phosphate to form ADP

B react with ATP synthase

C produce carbon dioxide

D pump hydrogen ions against a concentration gradient.

8. Which of the following is not correct?

A Oxytocin is produced by the pituitary

B Progesterone is a placental hormone

C Oestrogen is produced by the ovaries

D Follicle-stimulating hormone inhibits oestrogen.

9. The following table shows the percentage of men and women in a part of Britain who survived two different types of cancer for at least one year after diagnosis.

Type of cancer	Sex	1971–1975	1976–1980	1981–1985	1986–1990
oesophagus	males	12	13	17	22
	females	16	17	23	26
stomach	males	13	14	20	25
	females	15	16	22	27

Which of the following statements is not correct? Over the period of the study

A survival rates for men with stomach cancer have increased more than men with cancer of the oesophagus

B women have less chance of surviving stomach cancer compared with men

C fewer people survived cancer of the oesophagus than stomach cancer between 1986 and 1990

D in men, the greatest increase in survival was for stomach cancer between the periods 1976–1980 and 1981–1985.

10. The diagram below shows how a genetic human disease is passed on.

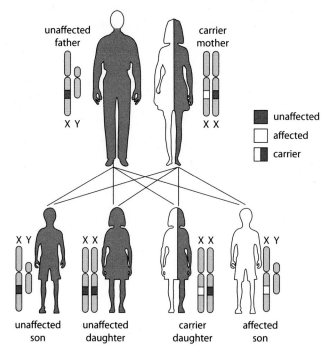

The allele for this disease is

A   dominant and not sex-linked

B   dominant and sex-linked

C   recessive and not sex-linked

D   recessive and sex-linked.

11. Which of the following is not correct? Tissue fluid

A   mostly ends up in capillaries

B   returns to the capillaries by osmosis and diffusion

C   allows the exchange of materials between cells and the blood

D   is forced out under pressure from capillaries and bathes body cells.

12. The graph below shows the changing heart rate, measured in beats per minute, for a person doing exercise for 6 minutes.

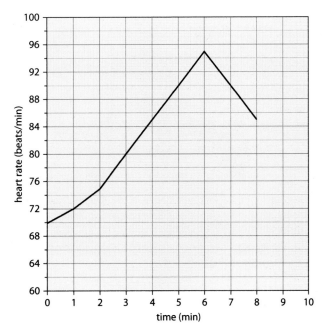

Assuming the rate of decrease in the heart rate remained the same, predict the likely heart rate (bpm) after 9 minutes.

A   80

B   78

C   76

D   75

13. High blood cholesterol level is

A   lowered if the ratio of low-density to high-density proteins is high

B   linked to the development of atherosclerosis

C   unaffected by exercise

D   unrelated to diet.

14. The following diagram shows the distribution of age groups in the population in Hong Kong in 2006.

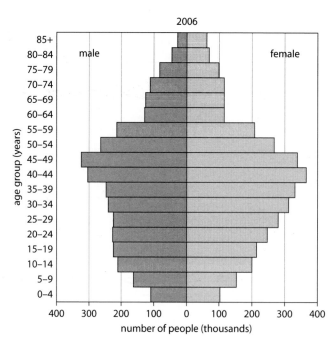

Which of the following conclusions is correct?

A   The numbers of males and females over 85 years is the same

B   There are more males in the 45–49 year-old range than any other range for males

C   Most females in the population are younger than 20 years

D   Males tend to live longer than females.

15. The type of memory involved in recalling personal facts and experiences is called

A   episodic

B   procedural

C   semantic

D   spatial.

Questions 16 and 17 refer to the following graph.

The graph shows the change in body temperature of two males, one aged 18 years and one aged 85 years, during exposure to an environmental air temperature of 15°C. No form of insulating clothing was worn during this experiment.

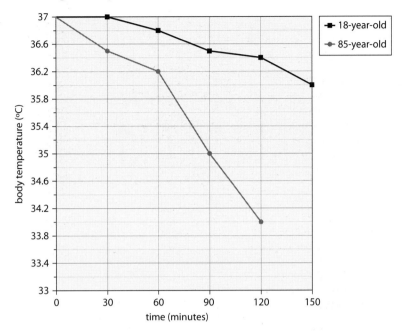

16. When the body temperature drops below 35°C, hypothermia sets in.

How long would it take the 85-year-old to develop hypothermia?

A   30 minutes

B   60 minutes

C   90 minutes

D   120 minutes.

17. If the drop in temperature of the 18-year-old continued at the rate it did from 120 to 150 minutes, predict his temperature after 180 minutes.

A   34·0°C

B   34·6°C

C   35·0°C

D   35·6°C

18. The diagram below represents a neural pathway. The arrows indicate the direction of nerve impulses.

The type of pathway shown can be described as

A   diverging

B   converging

C   sensory

D   reverberating.

19. Which of the following are examples of non-verbal communication?

A   Pitch of voice and nodding of head

B   Shouting and sitting upright in a class

C   Sticking out one's tongue and text-messaging

D   Sending an e-mail and waving hands.

20. The following stages are involved in the inflammation response:

1   Phagocytes accumulate and digest pathogens

2   Cytokines stimulate phagocytes to migrate to the infection site

3   Histamine causes blood vessels to dilate

4   Mast cells release histamine.

The correct sequence for these stages is

A   1, 2, 3, 4

B   2, 3, 4, 1

C   4, 3, 1, 2

D   4, 3, 2, 1

SECTION 2 – 80 marks

Attempt ALL questions.

It should be noted that question 13 contains a choice.

1. The following diagram shows the stages in the replication of DNA.

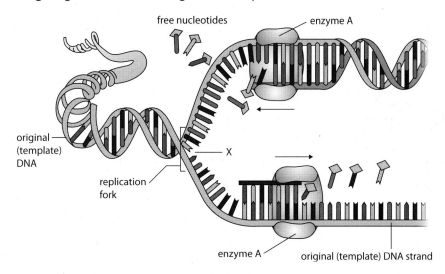

(a) Identify enzyme A. **1**

_____

(b) Using the numbers 1 and 2, identify **on the diagram** the strand of DNA which is replicated continuously and the strand of DNA which is replicated in fragments respectively. **2**

(c) Explain what is happening at point X. **1**

_____

_____

2. Patients with an infection in the lungs produce a great deal of mucus which may inhibit the effectiveness of an antibiotic.

A group of 50 patients with similar lung infections were divided into two groups. One group was treated with an antibiotic alone while the other had the same antibiotic plus an additional drug which helps to break up mucus and allow the antibiotic to be more effective.

Samples of mucus were collected every hour after the treatment was started and the concentration of antibiotic analysed. The higher the concentration of antibiotic in the sample, the more effective it is at treating the infection.

The results are shown in the table below.

Collection time (hours)	Concentration of antibiotic in mucus ($\mu g/cm^3$)	
	Antibiotic alone	Antibiotic plus additional drug
1	1·0	1·6
2	0·8	2·0
3	0·6	2·0
4	0·4	1·8
5	0·2	1·6
6	0·2	1·2
7	0·2	1·0
8	0	0·8
9	0	0·6
10	0	0·6

(a) State the length of time after which the antibiotic in the mucus sample is eight times higher in the presence of the additional drug compared to when the antibiotic is given alone.　**1**

_____ hours

(b) Give **two** differences in the results with the additional drug compared with the antibiotic alone.　**2**

1 _____

_____

2 _____

_____

2. **(continued)**

(c) State **two** design features of this investigation which could make the results unreliable. **2**

1 _____

_____

2 _____

_____

(d) Plot a line graph to show all the data in the table. **3**

(e) Based **only on the data given**, give **one** possible conclusion which can be made. **1**

_____

_____

3. The graph below shows one complete cycle of the polymerase chain reaction.

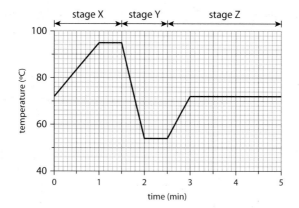

(a) State the function of the polymerase chain reaction. **1**

_____

_____

(b) Match each stage shown with the following descriptions: **2**

DNA strands separate: stage _____

Complementary DNA strands synthesised: stage _____

Primers bind: stage _____

(c) State the minimum change in temperature within one cycle of this reaction. **1**

Space for calculation

_____ °C

(d) Explain the function of a primer in DNA replication. **1**

_____

_____

3. (continued)

(e) A small sample of saliva containing one hundred molecules of DNA was found at a crime scene.

The sample was put through the polymerase chain reaction for 35 minutes.

Using data from the graph, calculate how many molecules of DNA would be synthesised after this time.   **1**

Space for calculation

_____ molecules

(f) Give **two** practical applications of the amplified DNA produced by polymerase chain reaction.   **2**

1 _____

_____

2 _____

_____

4. The diagram below shows some of the events associated with a cycle in reproduction.

Approximately halfway through the cycle, event A takes place.

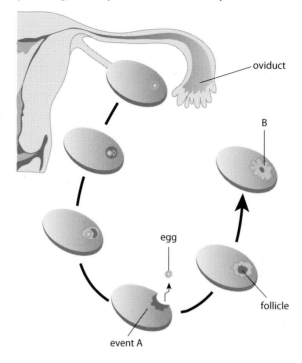

(a) (i)   State the term used to describe event A.   **1**

_____

(ii) Name the hormone which rises suddenly in the blood and triggers event A to
take place.   **1**

_____

(iii) State the gland which manufactures this hormone.   **1**

_____

(iv) State the name of structure B.   **1**

_____

(v) Name two hormones manufactured by structure B.   **2**

1 _____

2 _____

(b) Explain what is meant by *in vitro* fertilisation.   **1**

_____

_____

5. The following information refers to genetic diseases in humans.

Duchenne muscular dystrophy is sex-linked.

Huntington's disease is caused by an autosomal dominant allele.

Cystic fibrosis is an autosomal recessive disease.

The family tree below shows an inheritance pattern of one of these conditions.

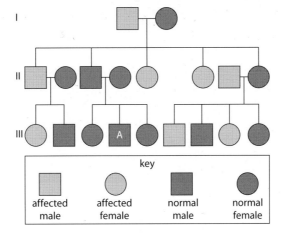

(a) (i) State which of the three conditions described is shown by this diagram.　　**1**

Condition: _____

(ii) Give **two** reasons for your answer.　　**2**

Reason 1 _____

_____

Reason 2 _____

_____

(b) Calculate the probability that individual A is heterozygous.　　**1**

Space for calculation

_____ %

5. (continued)

(c) The frequency of this disease is approximately 1 in 10,000. In a population of 5,000,000, approximately half are males and half are females. Calculate how many females are likely to have this disease. **1**

Space for calculation

_____

(d) Amniocentesis allows material from the developing fetus to be examined for abnormalities in the chromosomes.

State one other procedure which can be used for this type of antenatal testing. **1**

_____

6. (a) The graph below shows the average changes in female fertility and maximum heart rate in a sample of one hundred women over a period of 55 years.

The change in fertility is expressed as a percentage of the maximum potential.

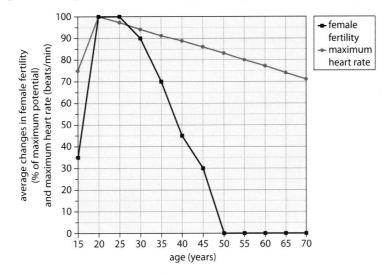

(i) State **two** conclusions which can be drawn from these results.   **2**

1 _____

_____

2 _____

_____

(ii) Assuming the rate of change for the maximum heart rate stays the same, predict the maximum heart rate for an eighty-year-old female.   **1**

_____ beats per minute

(iii) State the two ages which show the greatest change in female fertility.   **1**

_____ and _____ years

(iv) State how long the females in this sample have 100% fertility.   **1**

_____ years

(b) It is estimated that about 8% of women will stop menstruating by the age of forty years. In a population of two hundred women, 25% of them were found to be forty years or older.

Calculate how many will have experienced menopause.   **1**

Space for calculation

7. The following question relates to how quickly people react to a stimulus, measured in milliseconds (ms), if they have taken a caffeine drink before the trial.

One hundred people were randomly divided into two groups. Both groups were asked not to take any caffeinated drink for 24 hours before the experiment.

One group were given a caffeine drink while the other was given an identical drink except it had no caffeine in it.

Each person was given five 'trials', each of which consisted of fifty measurements of their reaction time. All the results, for each person and then for the whole group, were averaged as shown.

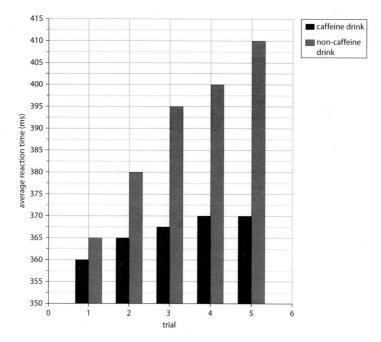

(a) The experimenter claimed that drinking caffeine lowered reaction time.

Give **two** reasons why this conclusion might not be correct.   **2**

1 _____

_____

2 _____

_____

(b) State **two** ways in which this experimental procedure could be improved.   **2**

1 _____

_____

2 _____

_____

8. The diagram below represents the organisation of the nervous system.

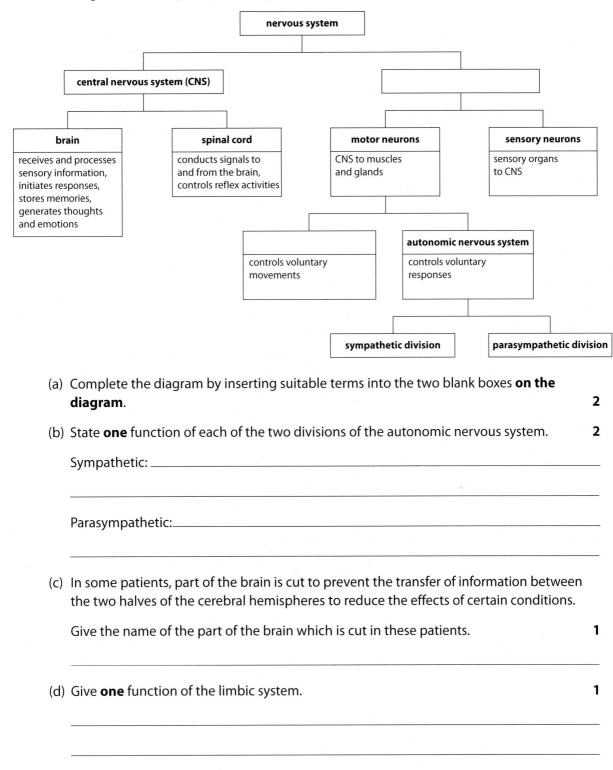

(a) Complete the diagram by inserting suitable terms into the two blank boxes **on the diagram**. **2**

(b) State **one** function of each of the two divisions of the autonomic nervous system. **2**

Sympathetic: _____

_____

Parasympathetic: _____

_____

(c) In some patients, part of the brain is cut to prevent the transfer of information between the two halves of the cerebral hemispheres to reduce the effects of certain conditions.

Give the name of the part of the brain which is cut in these patients. **1**

_____

(d) Give **one** function of the limbic system. **1**

_____

_____

9. Neurotransmitters are substances that allow nerve impulses to travel across synapses.

(a) Endorphins reduce pain and give a feeling of well-being.

Explain how these neurotransmitters function. **2**

_____

_____

_____

(b) Give **one** example of how the agonist dopamine can be used medically. **1**

_____

(c) Give **two** ways in which a neurotransmitter may normally be removed from the post-synaptic membrane. **2**

1 _____

_____

2 _____

_____

(d) Explain how a recreational drug might affect the transmission of nerve impulses across a synapse. **1**

_____

_____

10. An experiment was carried out in the late 1970s to study how learning a language could be improved. Groups of English speakers were presented with a series of words in another language, their meaning in English and an image whose name sounded like the foreign word.

Here is an example of the technique.

carta	*letter*	*cart*
Spanish word	Meaning in English	Image

One test group was given 120 Russian words to learn in this way while a second test group, a control, was given the same list but with no named images. The experimental group did 50% better when asked to recall the meanings immediately after the test compared with the control.

(a) Explain how this experiment suggests a way to improve our memory of foreign words and their meanings. **1**

_____

_____

(b) Give **one** other way of setting up a control experiment with the same groups of English speakers. **1**

_____

_____

(c) Suggest **one** limitation of learning a language in this way. **1**

_____

_____

11. (a) The graphs below show the response of a person, measured as antibody concentration, after exposure to two different antigens, A and B. Exposure to antigen A occurred 1·5 weeks after the start of the investigation and exposure to both antigens A and B simultaneously occurred 7·0 weeks after the start of the investigation.

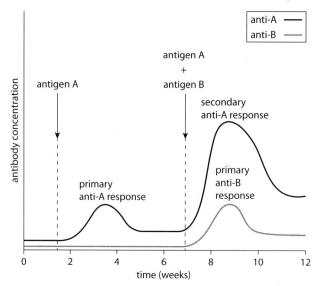

Explain the differences in the shapes of the graphs for the concentration of each of the two different antibodies, antibody A and antibody B.

(i) Antibody A: _____ **2**

_____

_____

_____

(ii) Antibody B: _____ **2**

_____

_____

_____

(b) Give **one** example of an auto-immune disease. **1**

_____

12. (a) Below are some control measures used to prevent the spread of infectious diseases:

- quarantine

- antisepsis

- vector control

Select **one** of these and explain how the control measure works. **3**

_____

_____

_____

_____

(b) The following statements describe how a disease can spread in a particular part of the world:

1. does not occur very often with no pattern and usually in isolated areas

2. frequency of occurrence is fairly constant in a particular area

3. an unusual increase in the frequency of occurrence above the expected level for a particular area

Match each statement by number to the following terms: **2**

Sporadic: _____

Epidemic: _____

Endemic: _____

(c) Give **one** way in which an individual can help prevent the spread of infectious diseases. **1**

_____

_____

_____

13. Answer either A or B in the space below

Labelled diagrams may be used where appropriate.

A   Describe fertility under the following headings:

    (i)   causes and treatment of female infertility;   **5**

    (ii) physical methods of contraception.   **5**

OR

B   Describe immunity under the following headings:

    (i)   non-specific defence mechanisms;   **5**

    (ii) design of vaccine clinical trials.   **5**

Mark scheme for Paper A

Section 1 (20 marks – 6 PS, 14 KU)

Question	Type	Mark	Expected response	Comment
1	KU	1	A	The main body tissue types are epithelial, connective, muscle and nerve tissue. The body organs are formed from a variety of these tissues.
2	PS	1	B	800 healthy cells at day 4. 360 cancer cells. 800 : 360 can be simplified by dividing both by 40. This gives 20 : 9.
3	PS	1	B	20% = thymine, 20% = adenine, so cytosine and guanine must be 30% each to take it to 100%. Therefore, thymine : cytosine = 20 : 30, simplified as 2 : 3.
4	KU	1	C	Structure of DNA – nucleotides contain deoxyribose sugar, phosphate and base. DNA has a sugar–phosphate backbone, complementary base pairing – adenine with thymine and guanine with cytosine. The two DNA strands are held together by hydrogen bonds and have an antiparallel structure, with deoxyribose and phosphate at 3' and 5' ends of each strand.
5	KU	1	B	Students often confuse introns and exons. Ensure you know the difference. The introns of the primary transcript of mRNA are non-coding and are removed in RNA splicing. The exons are coding regions and are joined together to form mature transcript.
6	KU	1	D	Arrays of DNA probes are used to detect the presence of specific sequences in samples of DNA. The probes are short single-stranded fragments of DNA that are complementary to a specific sequence. Fluorescent labelling allows detection.
7	KU	1	A	Ensure you know the two types for your exam and their differences. Slow twitch (type 1) muscle fibres contract more slowly, but can sustain contractions for longer and so are good for endurance activities. Fast twitch (type 2) muscle fibres contract more quickly, over short periods, so are good for bursts of activity.
8	KU	1	B	It's important to memorise the functions of all the reproductive hormones.
9	PS	1	B	Total decrease = 50 million/cm^3 50 / 60 years = 0·83 million/cm^3/year
10	KU	1	A	
11	KU	1	D	
12	PS	1	C	19 mm/500 (magnification) = 0·038 mm × 1000 (to change to micrometres) = 38 µm
13	PS	1	B	A is incorrect as people with type 2 diabetes have a decreased sensitivity to insulin after GBS compared to those who do not have type 2 diabetes. B is therefore the only correct conclusion, based on the data provided. C and D cannot be concluded from the data in the table alone.

14	KU	1	D	
15	KU	1	B	
16	KU	1	C	
17	KU	1	D	It's important to memorise the different roles of T and B lymphocytes.
18	KU	1	A	
19	KU	1	C	
20	PS	1	D	16% of children with asthma ate oily fish, which is less than 30% of those without asthma, so A is true. 28% of children with asthma did not eat fish, that's 72% who do, whereas 16% of children without asthma did not eat fish, that's 84% who do. Therefore, B is also true. 56% of children with asthma ate non-oily fish compared to 54% in children without asthma. This a 2% difference, so C is also true. D is wrong because it was the children with asthma who had 28% eating oily fish.

Section 2 (80 marks – 24 PS, 56 KU)

Question			Type	Mark	Expected response	Comment
1	a		KU	1	Germline.	Germline cells divide by mitosis to produce more germline cells or by meiosis to produce haploid gametes.
	b		KU	1	Embryonic and tissue (adult) stem cells.	
	c		KU	1	Tissue stem cells are more limited in the cells they can make, whereas embryonic can make all differentiated body cells. Tissue are multipotent whereas embryonic are pluripotent. Tissue stem cells found in the tissue that they replace, whereas embryonic found in early embryo.	Tissue (adult) stem cells are involved in the growth, repair and renewal of the cells found in that tissue. They are multipotent. The cells of the early embryo can make all of the differentiated cell types of the body. They are pluripotent. When grown in the lab scientists call these embryonic stem cells.
	d		KU	1	The process by which a cell develops more specialised functions by expressing the genes characteristic for that type of cell.	
	e		KU	3	Reference to the repair of damaged or diseased organs or tissues. (1) Named example. (1) Used as model cells to study how diseases develop or for drug testing. (1) The ethical issues of stem cell use and the regulation of their use. Mention of the use of embryos needed to gain mark. (1)	
	f		KU	1	They can spread through the body to form secondary tumours.	
2	a		KU	1	Nucleus.	
	b		KU	1	RNA polymerase.	
	c		KU	2	Moves along DNA unwinding and unzipping the double helix. (1) Synthesising a primary transcript of RNA by complementary base pairing. (1)	
	d		KU	1	RNA is single stranded while DNA is double stranded. RNA has a ribose sugar while DNA has deoxyribose sugar.	RNA is single stranded, contains uracil instead of thymine and ribose instead of deoxyribose sugar.

	e		KU	3	RNA splicing. (1) The introns of the primary transcript of mRNA are non-coding and are removed in RNA splicing. (1) The exons are coding regions and are joined together to form mature transcript. (1)	Different proteins can be expressed from one gene as a result of alternative RNA splicing and post-translational modification.
3	a		KU	2	X – ATP Y – ADP Z – Pi	
	b		KU	1	Catabolic.	Anabolic pathways require energy and involve biosynthetic processes. Catabolic pathways release energy and involve the breakdown of molecules.
	c		KU	1	Final electron acceptor (combines with hydrogen ions and electrons to form water).	
	d		KU	1	Peptide.	
4	a		PS	1	Temperature of the test tube contents. Concentration of yeast/substrate/methylene blue.	
	b		PS	1	The experiment repeated several times for each substrate. OR The experiment repeated and an average calculated.	
	c		PS	1	To allow time for the yeast to start respiring/respiration rate to stabilise.	
	d		PS	2	Control. (1) To ensure that any change in results is due to the variable that is altered and to allow a valid comparison with the variable being investigated. (1)	Control experiments are set up to ensure that any change in results is due to the variable that is altered. Control experiments have to be identical to the original experiment but minus the altered variable. Suitable controls may be to replace organisms with the same mass or volume of a non-living material, the same mass of a boiled tissue or the same volume of water rather than the given solution.

	e	PS	2	Correct scales and labels on axes. (1) Points correctly plotted and line drawn. (1)	Remove one mark if urea concentration is plotted on the vertical axis. Remove 1 mark if less than half the graph paper is used.
	f	PS	1	As the glucose concentration increases, the time taken for the methylene blue to become colourless decreases. OR Inverse.	
5	a	KU	2	X – central lumen Y – connective tissue	
	b	KU	1	Stretch and recoil to accommodate the surge of blood after each contraction of the heart.	
	c	KU	2	Contract/vasoconstriction. (1) Blood vessels supplying the capillaries in the skin contract which causes an increase in blood pressure. (1) OR Relax/vasodilation. (1) Blood vessels supplying the capillaries in the skin relax which causes a decrease in blood pressure. (1)	
	d	KU	3	Central lumen – arteries are narrower than vein OR inverse. Muscular wall – arteries much thicker than veins OR inverse. Valves – veins contain valves while arteries do not. Direction of blood flow – arteries take blood away from heart and veins return blood to heart. Elasticity – artery walls more elastic than vein or inverse.	Arteries have an outer layer of connective tissue containing elastic fibres and a middle layer containing smooth muscle with more elastic fibres. The elastic walls of the arteries stretch and recoil to accommodate the surge of blood after each contraction of the heart. The smooth muscle can contract or relax causing vasoconstriction or vasodilation to control blood flow. Veins have an outer layer of connective tissue containing elastic fibres but a much thinner muscular wall than arteries. Function of valves.

6	a		KU	1	One number is the systolic pressure and the other is the diastolic pressure which represent the blood pressure changes in the aorta during the cardiac cycle.	The blood starts to flow (detected by a pulse) at systolic pressure. The blood flows freely through the artery (and a pulse is not detected) at diastolic pressure.
	b		KU	1	A typical reading for a young adult is 120/70 mmHg.	Accept +/− 5 on either value.
	c		KU	1	Coronary heart disease. OR Others, including stroke or diabetes.	
7	a		PS	2	In group 1 most said 13 – this is probably due to them counting and therefore more likely to perceive the stimulus as a number. (1) In group 2 most said B – this is probably due to them reciting letters and therefore they are more likely to perceive the stimulus as a letter. (1)	
	b		PS	1	They only conducted the experiment once with each group (so an average was not calculated).	
	c		KU	1	Factors which influence the way a stimulus is perceived.	
	d		KU	2	Past experience, context and expectation.	3 correct = 2 marks 1 or 2 correct = 1 mark
8	a		KU	1	Individual C – XgY Individual D – XGXg	Must use letter g. Accept – XgXG.
	b		KU	1	The allele is carried on the X chromosome and the father/individual A passes the Y chromosome to his son. OR The allele is carried on the X chromosome and the father/individual A does not pass it to his son. OR The allele is carried on the X chromosome and the father/individual A passes the X chromosome to his daughter.	Accept condition/it/gene for allele.
	c		PS	1	50.	
	d		KU	1	Mutation.	
	e		KU	2	The sequence/order of amino acids is changed (from that point). (1) OR Different amino acids (in enzyme). (1) The shape/active site of the enzyme will change. (1)	Not – structure of enzyme changes.

	f		PS	1	250.	
	g		KU	1	Genetic screening/genetic counselling/ pedigree chart analysis.	
	h		KU	1	Pre-implantation Genetic Diagnosis/PGD/ pre-implantation genetic screening.	
9	a		PS	1	USA.	China = 110 / 1375 × 100 = 8% India = 70 / 1311 × 100 = 5·3% USA = 30 / 320 × 100 = 9·4% Brazil = 14 / 208 × 100 = 6·7%
	b		PS	1	6:7	
	c		KU	2	(see table below)	

Statement	Type 1 diabetes	Type 2 diabetes
Usually occurs in childhood	√	
Typically develops later in life		√
Occurs mainly in overweight individuals		√
Unable to produce insulin	√	
Can be treated with regular doses of insulin	√	
Individuals produce insulin but their cells are less sensitive to it		√
Blood glucose levels rise rapidly after a meal	√	√

10	a		PS	1	11.	
	b		PS	1	As children age from 6 to 14, their memory span for numbers increases. After 14, there is a small decrease in memory span.	
	c		PS	1	The sequence of numbers/gender ratio/the method of recall/the conditions within the room such as temperature, light intensity, noise levels, proximity to other children, etc. Any other suitable answer.	Any one.

	d		PS	1	Easier sequence of numbers. Students in the 12–14 age group may have had prior experience or more practice than children in the 15–17 age group. Small sample size in all groups means that the results are not representative of all children of that age.	
	e		PS	1	Increase the sample size for each group. Repeat the experiment for each age group several times and take an average.	
	f		KU	1	Social facilitation.	
	g		KU	1	Reinforcement.	
11	a		PS	1	86 bpm.	$60 / 0.7 = 85.7$
	b		PS	1	5022 cm^3/min.	$148 - 55 = 93$ $93 \times 54 = 5022$ cm^3/min Cardiac output is determined by heart rate and stroke volume $(CO = HR \times SV)$.
	c		PS	2	Closed Open Open Closed.	
12	A	i	KU	3	1. Pituitary gland is stimulated 2. to release follicle-stimulating hormone (FSH) 3. luteinising hormone (LH) 4. or interstitial cell-stimulating hormone (ICSH) 5. by a releaser hormone produced in the hypothalamus.	Any 3 from 5.
	A	ii	KU	3	1. FSH promotes sperm production 2. and ICSH stimulates the production of testosterone. 3. Testosterone also stimulates sperm production 4. and activates the prostate gland and seminal vesicles. 5. Negative feedback control of testosterone by FSH and ICSH.	Any 3 from 5.

	A	iii	KU	4	1. FSH stimulates the development of a follicle and the production of oestrogen by the follicle in the follicular phase.	Any 4 from 7.
					2. Oestrogen stimulates proliferation of the endometrium preparing it for implantation and affects the consistency of cervical mucus making it more easily penetrated by sperm.	
					3. Peak levels of oestrogen stimulate a surge in the secretion of LH which triggers ovulation.	
					4. In the luteal phase the follicle develops into a corpus luteum and secretes progesterone.	
					5. Progesterone promotes further development and vascularisation of the endometrium, preparing it to receive a blastocyst if fertilisation occurs.	
					6. The negative feedback effect of the ovarian hormones on the pituitary gland and the secretion of FSH and LH prevent further follicles from developing.	
					7. The lack of LH leads to degeneration of the corpus luteum with a subsequent drop in progesterone levels leading to menstruation.	
	B	i	KU	5	1. Infectious diseases caused by pathogens,	Any 5 from 7.
					2. transmitted by direct physical contact, water, food, body fluids, inhaled air or vector organisms	
					3. controlled by quarantine,	
					4. antisepsis,	
					5. individual responsibility (good hygiene, care in sexual health and appropriate storage/handling of food),	
					6. community responsibility (quality water supply, safe food webs and appropriate waste disposal systems)	
					7. and vector control.	

	B	ii	KU	5	1. Description of spread to include sporadic (occasional occurrence),	Any 5 from 7.
					2. endemic (regular cases occurring in an area),	
					3. epidemic (unusually high number of cases in an area)	
					4. or pandemic (a global epidemic).	
					5. Control measures to include preventing transmission,	
					6. drug therapy,	
					7. immunisation or a combination of these.	

Mark scheme for Paper B

Section 1 (20 marks – 6 PS, 14 KU)

Question	Type	Mark	Expected response	Comment
1	KU	1	D	RNA contains uracil (U), not thymine (T), so this rules out A and C. RNA is single stranded, so this rules out B and C. RNA contains ribose sugar. This rules out A and B.
2	KU	1	A	
3	PS	1	D	
4	KU	1	D	
5	PS	1	D	60 mm / 100 = 0·6 mm 0·6 × 1000 = 600 micrometres
6	KU	1	C	
7	KU	1	D	Ensure you learn all the reproductive hormones and their functions. Use flashcards to help you and get someone to quiz you on them.
8	KU	1	B	
9	KU	1	A	
10	KU	1	C	As well as learning the names of the phases of the cardiac cycle, you should be familiar with diagrams of the heart in each phase.
11	KU	1	C	
12	PS	1	D	
13	PS	1	A	The non-diabetic people are a control group. Controls allow a valid comparison to ensure the result is due to the variable being investigated.
14	KU	1	A	
15	KU	1	D	
16	KU	1	C	
17	KU	1	A	
18	KU	1	B	
19	PS	1	D	
20	PS	1	D	Graphs like this are hard to read. Take care to look at both scales and be sure which line refers to which scale. Use a pencil to draw lines at appropriate places on your graph to help you identify the appropriate data.

For Question 3, the Comment cell contains the following table:

	1	2	3	4	5
DNA molecules	1,000	2,000	4,000	8,000	16,000

Each cycle doubles the number of DNA molecules. Use a simple flow diagram or table like this to double the number each time for the correct number of cycles.

Section 2 (80 marks – 26 PS, 54 KU)

Question			Type	Mark	Expected response	Comment
1	a		KU	1	Hydrogen.	
	b		KU	1	Nucleotide.	
	c		KU	1	Deoxyribose sugar.	
	d		KU	2	Have an antiparallel structure (1) with deoxyribose and phosphate at 3′ and 5′ ends of each strand. (1)	
	e		KU	1	DNA polymerase.	Don't confuse with RNA polymerase. It's important to memorise the functions of both DNA polymerase and ligase in DNA replication.
	f		KU	2	ATP. (1) Enzymes (DNA polymerase and DNA ligase). (1) Primers. (1)	Any 2 of 3.
	g		KU	1	Ensures new cells have the same chromosomes/DNA/genes/genetic information/are genetically identical/have a complete copy of the genome. OR To maintain the correct quantity/number complement of DNA/chromosomes.	
2	a		KU	1	Bioinformatics.	
	b		KU	1	To provide information on evolutionary relationships and origins.	
	c		KU	1	• Shift the emphasis in medicine from reaction to prevention. • Predict susceptibility to disease. • Improve disease detection. • Pre-empt disease progression. • Customise disease-prevention strategies. • Prescribe more effective drugs. • Avoid prescribing drugs with predictable side effects. • Reduce the time, cost and failure rate of pharmaceutical clinical trials.	Any one or any other suitable.
	d		KU	1	Short single-stranded fragments of DNA that are complementary to a specific sequence.	
3	a		KU	1	Creatine phosphate.	

	b		KU	1	Pyruvate is converted to lactic acid.	This conversion involves the transfer of hydrogen from the NADH produced during glycolysis to pyruvic acid to produce lactic acid.
	c		KU	2	The volume of oxygen needed to oxidise lactic acid to carbon dioxide and water. (1) Causes us to breathe deeply and quickly for a while after exercise. (1)	
	d		KU	2	Cell B. (1) Contains many more mitochondria which allows larger quantities of ATP to be produced for the prolonged activity. (1)	Slow twitch muscle fibres are good for endurance activities like long distance running, cycling or cross-country skiing. Slow twitch muscle fibres rely on aerobic respiration to generate ATP and have many mitochondria, a large blood supply and a high concentration of the oxygen-storing protein myoglobin.
	e		KU	3	Slow twitch (type 1). Fast twitch (type 2). Difference: Slow twitch (type 1) muscle fibres contract more slowly, but can sustain contractions for longer and so are good for endurance activities. Fast twitch (type 2) muscle fibres contract more quickly, over short periods, so are good for bursts of activity.	Any one difference. Difference must be comparative.
4	a		PS	2	As the age increases from <10 to 60–69, the incidence rate of type 2 diabetes increases from 0 to 733. (1) After 69 the incidence rate decreases from 733 to 649. (1)	
	b		KU	1	Type 2 diabetes typically develops later in life (due to insulin resistance).	
	c		PS	1	14·85 or 15%	% change = change / original $\times$ 100 change = 116 – 101 = 15 15 / 101 $\times$ 100 = 14·85%

	d		PS	1	683	2012 = 833, 2013 = 783, 2014 = 733 A decrease of 50 each year, so 733 − 50 = 683
5	a		PS	1	As the age of women increases (from under 34 to over 42), the percentage of live births per treatment cycle decreases (from 27·6 to 5·0).	
	b		PS	1	63·77 or 63·8%	% change = change / original × 100 change = 27·6 − 10 = 17·6 17·6 / 27·6 × 100 = 63·77
	c		KU	1	Physical blockage in reproductive tracts. Low sperm count/unable to produce sperm/sperm have low motility. Inability to stimulate ovulation/eggs not produced/egg immature. Failure of zygote to implant in uterus. Any other suitable.	Any one.
	d		KU	1	Artificial insemination. Intra-cytoplasmic sperm injection (ICSI). Stimulating ovulation using drugs.	Any one.
	e		KU	1	Physical methods could include barrier methods, avoiding fertile periods, intra-uterine devices and sterilisation procedures. Chemical contraceptives including pills – some prevent implantation ('morning-after pills') or cause thickening of cervical mucus ('progesterone-only pill').	Any one.
6	a		PS	1	The population of the country is likely to have been different each year.	
	b		KU	3	HIV attacks lymphocytes. (1) HIV is the major cause of acquired immunodeficiency in adults. (1) Tuberculosis (TB) survives within phagocytes and avoids immune detection. (1)	
7	a		PS	1	Group – rewarded.	Rewarded: 395 − 284 = 111 111 / 395 × 100 = 28.1% Not rewarded: 303 − 275 = 28 28 / 303 × 100 = 9·2%
	b		PS	1	Student 5.	

	c	PS	1	The average time to complete the Sudoku puzzle decreased between attempts 1 and 10 in both groups. OR The average time to complete the Sudoku puzzle decreased initially but then increased on attempt 10 in both groups.	
	d	PS	1	The puzzles given to each student should be the same. The puzzles should be given in same order. Students should not see the puzzles before the time starts. Students should not interact between each puzzle. Any other acceptable.	Any one.
	e	PS	1	Environmental conditions such as temperature, light intensity, noise levels.	Any one.
	f	PS	1	Repeat the experiment exactly as it is. Repeat with a greater number of students in each group.	Any one.
	g	PS	2	Correct scales and labels on axes. Bars are correctly drawn.	Remove 1 mark if less than half the graph paper is used.
	h	PS	1	Students who are rewarded for successfully completing a puzzle show a greater decrease in time to complete subsequent puzzles compared with those who are not rewarded.	
8	a	PS	1	17	$70 - 53 = 17$
	b	PS	1	13·6 or 14%	% change = change / original $\times$ 100 change = $81 - 70 = 11$ $11 / 81 \times 100 = 13\cdot6$
	c	PS	1	The starting heart rates for the men and women were different so it would not be a valid comparison.	
	d	KU	1	Stroke volume increased.	Cardiac output is determined by heart rate and stroke volume (CO = HR $\times$ SV). So if HR decreased and SV increased, then CO could stay the same.
9	a	PS	2	Correct scales and labels on axes. Bars correctly drawn.	Remove 1 mark if less than half the graph paper is used.

	b		PS	1	1	8 letters = 7, 9 letters = 5, 10 letters = 3 This is a drop of 2 students each time 1 letter is added.
	c		PS	1	As the number of letters in the sequence increased, the number of students who recalled all the letters in the correct order decreased.	
	d		PS	1	5:4	40:32 – Both can be divided by 8. 5:4
	e		PS	1	To ensure a consistent speed and tone so that all students experienced the same experimental conditions and variables.	
	f		PS	1	To increase the sample size and reduce the magnitude of experimental error.	
	g		KU	1	Chunking.	
	h		KU	1	Displacement.	
10	a		KU	1	Placebo.	
	b		KU	1	To ensure valid comparisons. To ensure any positive effect is due to the drug.	Any one.
	c		KU	1	Double-blind trial.	It's important to learn all the terms and the reasons for experimental procedures and drug testing.
	d		KU	1	To ensure the subjects were randomised and to eliminate bias.	
	e		KU	1	To reduce the magnitude of experimental error. To possibly increase the statistical significance of the results.	Any one.
11	a		KU	1	Immune surveillance.	
	b		KU	2	Lymphocytes have a single type of membrane receptor specific for one antigen. (1) Antigen binding leads to repeated lymphocyte division resulting in a clonal population of lymphocytes. (1)	
	c		KU	2	T lymphocytes have specific surface proteins that allow them to distinguish between the surface molecules of the body's own cells and cells with foreign molecules on their surface. (1) Immune system regulation failure leads to T lymphocytes immune response to self antigens. (1)	

	d		KU	2	Hypersensitive B lymphocyte response (1) to an antigen that is normally harmless. (1)	
12	A	i	KU	5	1. Neurotransmitters relay messages from nerve to nerve within and outwith the brain. 2. Neurons connect with other neurons, muscle fibres and endocrine glands at a synaptic cleft. 3. Neurotransmitters are stored in vesicles 4. and released into the cleft on arrival of an impulse. 5. They diffuse across the cleft and bind to receptors on nerve endings. 6. Receptors determine whether the signal is excitatory or inhibitory. 7. Synapses can filter out weak stimuli arising from insufficient secretion of neurotransmitters. 8. Summation of a series of weak stimuli can trigger enough neurotransmitters to fire an impulse.	Any 5 from 8.

A	ii	KU	5	1. Endorphins are neurotransmitters that stimulate neurons involved in reducing the intensity of pain. 2. Increased levels are also connected with euphoric feelings, appetite modulation and release of sex hormones. 3. Endorphin production increases in response to severe injury, prolonged and continuous exercise, stress and certain foods. 4. Dopamine induces the feeling of pleasure and reinforces particular behaviour in the reward pathway. 5. Neurotransmitter-related disorders and their treatment. 6. Agonists bind to and stimulate receptors mimicking the neurotransmitter. 7. Antagonists bind to specific receptors blocking the action of the neurotransmitter. 8. Other drugs inhibit the enzymes which degrade neurotransmitters or inhibit re-uptake.	Any 5 from 8.
B	i	KU	5	1. During diastole 2. blood returning to the atria flows into the ventricles. 3. Atrial systole 4. transfers the remainder of the blood through the atrio-ventricular (AV) valves to the ventricles. 5. Ventricular systole 6. closes the AV valves 7. and pumps the blood out through the semi lunar (SL) valves 8. to the aorta and pulmonary artery. 9. In diastole the higher pressure in the arteries closes the SL valves.	Any 5 from 9.

	B	ii	KU	5			Any 5 from 9.
					1.	The heart beat originates in the heart itself but is regulated by both nervous and hormonal control.	
					2.	The auto-rhythmic cells of the sino-atrial node (SAN) or pacemaker set the rate at which cardiac muscle cells contract.	
					3.	The timing of cardiac cells contracting is controlled by the impulse from the SAN spreading through the atria	
					4.	and then travelling to the atrio-ventricular node (AVN)	
					5.	and then through the ventricles.	
					6.	The medulla regulates the rate of the SAN	
					7.	through the antagonistic action of the autonomic nervous system (ANS).	
					8.	Sympathetic accelerator nerves release nor-adrenaline (nor-epinephrine)	
					9.	and slowing parasympathetic nerves release acetylcholine.	

Mark scheme for Paper C

Section 1 (20 marks – 7 PS, 13 KU)

Question	Type	Mark	Expected response	Comment
1	KU	1	B	20% will be guanine which means thymine and adenine each make up 30% of the molecule.
2	KU	1	C	Introns are non-coding and removed in the process of RNA splicing from primary to mature transcript.
3	KU	1	C	Each type of gene mutation, missense, nonsense and splice-site has a different effect on the translation process.
4	KU	1	D	All the children have a band match with this father.
5	KU	1	D	Competitive inhibitors are similarly shaped to the substrate and block the active site of the enzyme.
6	PS	1	B	
7	PS	1	A	By adding an equal volume of water to an equal volume of the stock solution, the technician is effectively diluting it by a factor of 2 from 1·8% to 0·9% which is what is required.
8	PS	1	C	To answer this question, you need to find two points on the graph where the trace starts to repeat itself. You can choose, for example, time 0 s and time 0·8 s or the two peaks. The time will be 0·8 s which gives a calculated value for the heart rate of $(10 \times 60) \div 8 = 75$ beats per minute.
9	KU	1	B	
10	KU	1	A	
11	KU	1	A	
12	KU	1	B	
13	PS	1	B	You need to check each statement here. After 30 minutes John's blood glucose level was 140 mg/100 cm^3 which is not twice his original value of 85 mg/100 cm^3 so A is incorrect. James' blood glucose level dropped from 165 to 100 = 65 mg/100 cm^3 which represents a percentage drop of $(65 \div 165) \times 100 = 39\%$, which is very close to the value of 40% making B correct. After 60 minutes James' blood glucose level was 10 mg/100 cm^3 higher than John's which represents a percentage of $(10 \div 90) \times 100 = 11\%$ not 20% so C is incorrect. John returned to his base level of 85 mg/100 cm^3 within 90 minutes whereas James was still higher than his original after the same time so D is incorrect.

14	KU	1	B	
15	KU	1	C	
16	PS	1	C	Generally, practising improves performance for most human activities. You need to be careful not to make any conclusions which are not supported by the evidence. The improvement was not uniform over the trial period since the graph is not linear so A is wrong. The time taken to perform the task does seem to decrease with practice so B is wrong. Most improvement was shown at the early stages, with the first 10 or so trials, after which time the improvement starts to level out so C is correct. You cannot extrapolate here (go further) with the data since you don't actually know what would happen with further trials so D is wrong.
17	KU	1	C	The boy has developed a fear of all snakes based on his experience with one. This is called generalisation, the ability to respond to related stimuli in the same way.
18	KU	1	B	Apoptosis is a highly regulated sequence of events which results in cell death and is a non-specific defence mechanism.
19	PS	1	C	Be careful here not to take the first temperature reading at five minutes and watch the scaling. The highest temperature recorded was 37·05°C and the lowest 36·78°C so the range is 0·27°C.
20	PS	1	D	Assuming no further procedure, the body temperature will remain very close to 36·80°C.

Section 2 (80 marks – 24 PS, 56 KU)

Question			Type	Mark	Expected response	Comment
1	a		KU	1	Normal cells start to divide uncontrollably.	Cells which have turned cancerous have lost the ability to regulate mitosis.
	b		KU	1	Tumour.	
	c		KU	1	Abnormal cells may not respond to regulatory signals.	Cells often communicate with each other via proteins which can stimulate or inhibit cell division.
	d		KU	1	Spread to other parts of the body.	Cells normally stick to each other but this may break down in cancerous cells allowing them to escape into the bloodstream to set up secondary growths.
2	a	i	KU	1	A nucleotide/base may be changed.	This is an example of a mutation.
		ii	KU	1	Protein/enzyme coded for will have an altered amino acid sequence.	
	b		KU	1	Early diagnosis means treatment can be put in place before any serious effects of the condition can occur.	
	c		KU	1	Person would be put on a low phenylalanine diet.	This is normally a lifetime commitment.
	d		KU	2	Autosomal: disease is not linked to sex of either parent/occurs equally often in males and females. (1) Recessive: disease does not appear in every generation/two unaffected parents can have an affected child. (1)	
3	a		PS	1	5.7	
	b		PS	2	Each axis correctly labelled. (1) Data plotted correctly. (1)	Remember to label each axis correctly and include units where appropriate (here simply 'average activity'). Make sure to use more than 50% of the grid.

	c	i	PS	1	By repeating the experiment.	Repeating experiments improves the reliability but remember it does not improve the accuracy.
		ii	PS	2	Constant temperature. Same time in blender. Same method of measuring volumes of oxygen given off.	Any two marks.
		iii	PS	1	Use boiled tissue.	Catalase is an enzyme and a common control for enzyme experiments to prove enzyme function is to boil the tissue/enzyme source to denature the protein structure of the enzyme.
		iv	PS	1	Catalase activity varies in different plant tissues/parsnip has a relatively high catalase activity compared with radish which has a relatively low catalase activity.	
		v	PS	1	1:3	Remember to preserve the order asked in the question.
4	a		KU	2	Light emission is easy to measure. (1) Measurement can be made very accurately/is objective. (1)	
	b		KU	3	Cells are functioning very efficiently. (1) The more ATP produced, the greater the quantity of luciferin converted into oxyluciferin. (1) The more oxyluciferin present, the greater the light emission. (1)	
	c		KU	1	It is very sensitive.	
5	a		KU	1	Phosphorylation.	The addition of a phosphate group is a very common process in cell biochemistry.
	b		KU	1	Pyruvic acid/pyruvate.	
	c		KU	1	Dehydrogenase.	This is a generic name for any enzyme capable of removing hydrogen from a molecule.
6	a		KU	1	Underline endurance sports and marathon running.	
	b		KU	1	Slow twitch.	
	c		KU	1	Fast twitch.	
	d		KU	1	Fast twitch.	
7	a		PS	1	315	

	b		PS	1	Ensure sex/age/level of fitness of group sample was same.	
	c		PS	1	High jumpers.	
	d		PS	1	Cycling.	
	e		PS	1	High jumpers.	
8	a		KU	1	There are three copies of chromosome (21) instead of two.	
	b		KU	1	Karyotype.	
	c	i	KU	1	Male.	
		ii	KU	1	Fetus.	
9	a		KU	1	Prenatal.	
	b		KU	4	Sample of amniotic fluid is taken. (1) Fluid contains cells from developing baby. (1) Cells can be examined for abnormal chromosome complement. (1) Increased chance of miscarriage/ possible risk of infection/injury to developing baby. (1)	
10	a		PS	1	4·5 months/18 weeks.	
	b		PS	1	1,150	
	c		PS	1	20%	
	d		PS	1	Birth mass.	
	e		PS	2	More babies are born overall. Birth mass is greater. Circumference of the head is greater. Length of baby is greater.	Any two.
	f		PS	2	Improve health education. Target pregnant women. Advertise in antenatal clinics. Involve family doctors in raising awareness.	Any two.
11	a		KU	1	Central nervous system.	The nervous system in humans is very complex and is best learned using diagrams and flow charts. Flash cards are an ideal way of reinforcing the learning of the terms.
	b		KU	1	Autonomic system.	This controls activities below the level of consciousness such as breathing, digestion and heart beat.
	c		KU	1	Motor nerve cells/motor neurons.	

	d		KU	1	Receptor; dendrites; cell body; axon; effector.	
12	a		KU	1	Joins the two cerebral hemispheres together/acts as a bridge between the two halves of the cerebrum.	
	b	i	PS	1	Chicken leg.	This is a difficult question on split-brain function. When the corpus callosum is cut, the two hemispheres can't communicate with each other properly and so, while the person perceives both images, they can only 'speak' the image of the chicken leg which the left hemisphere has seen because that is also where the speech centre is located.
		ii	PS	1	Snow-shovel.	The person's right hemisphere has 'seen' the snow-scene. He can't speak the words because the right hemisphere does not have language function but he can indicate what he saw with his left hand whose function is controlled by the right hemisphere.
13	a		KU	1	Imitation.	
	b		KU	1	Learning to use a machine/perform a new task/perform sport.	There are a variety of possible answers here but the main thing is to show that humans can learn by imitation in adult tasks by watching people who are technically accomplished in those tasks and then trying to imitate them.
	c		KU	1	Verbal praising/giving a sweet/smiling when task is performed correctly.	There are many examples of positive reinforcement. Three are given here as ways to positively enhance someone's learning.
			KU	1	Shouting/angry expression on face/disapproving tone of voice.	Similarly, there are many examples of negative reinforcement. Three are given here as ways to modify someone's behaviour by using a negative strategy of reinforcement.
	d		KU	1	Lays foundation of future formation of stable relationships.	
	e		KU	1	Increased performance when in a group.	

	f		KU	3	Deindividuation occurs in a crowd when people lose their sense of personal identity. (1) In a crowd, people may be more aggressive because they are less accountable for their individual actions. (1) Under video recording people know their individual actions may be logged and so they are less liable to acts of violent behaviour. (1)	
14	A	i	KU	5	1. Organs suspended in the scrotum 2. produce male gametes/sperm 3. in the long, coiled seminiferous tubules 4. hundreds of which are present in each testis 5. and stored until used. 6. Also produce testosterone 7. in the interstitial cells between the seminiferous tubules 8. under the influence of interstitial cell-stimulating hormone 9. produced by the pituitary. 10. Testosterone contributes to the onset of puberty.	Any 5 from 10.

	A	ii	KU	5	1. Prostate gland is a solid chestnut-shaped organ situated under the bladder 2. which adds a milky alkaline fluid to the seminal fluid 3. a creamy, sticky fluid forced out of the penis during ejaculation 4. secreted by the prostate gland and the seminal vesicles 5. contains nutrients such as fructose required by the sperm for energy. 6. Alkaline fluid helps neutralise any traces of acidic urine in the urethra 7. and the female vaginal tract 8. since sperm cannot swim well in an acidic environment. 9. Seminal vesicles are two small, elongated sacs situated on the prostate gland 10. at the point where the sperm duct passes through to join the urethra. 11. first accessory glands to add fluid to the sperm 12. as it travels along the sperm ducts.	Any 5 from 12.

14	B	i	KU	5	1.	Active immunity is the production of antibodies to a specific antigen by host.	Any 5 from 9, of which maximum of 2 marks for examples of different types of vaccines.
					2.	How vaccine is administered is critical to its success.	Suitably drawn and labelled diagrams may be used if the information is relevant.
					3.	Vaccines may be administered by mouth/nasal spray/injection.	
					4.	Some B lymphocytes act as memory cells while others produce antibodies.	
					5.	Vaccine is used to provoke primary immune response while not causing symptoms.	
					6.	Vaccine may be an inactivated form of the antigen such as against flu and cholera.	
					7.	Vaccine may be live or weakened form of antigen such as against measles and mumps.	
					8.	Vaccine may be an inactivated toxin produced by a microorganism such as tetanus and diphtheria.	
					9.	Vaccine may be a small piece of the antigen such as against hepatitis B.	

	B	ii	KU	5	1.	Occurs when a large percentage of the population has been immunised against an infection.	Any 5 from these 8.
							Suitably drawn and labelled diagrams may gain marks for information not in the text.
					2.	This provides a level of immunity towards those members of the population who have not been immunised.	
					3.	The higher the percentage immunised, the lower the chances of those non-immunised contacting an infected individual.	
					4.	Infectious disease is thus much less likely to spread.	
					5.	Helps protect at risk individuals such as elderly or infirm or newborn babies who have not been immunised.	
					6.	If the percentage immunised falls below some threshold value, the infectious disease can spread much more easily.	
					7.	The threshold value is influenced by the type of infectious disease and how efficient the vaccine is.	
					8.	Less wealthy countries may not be able to immunise on a largescale so reducing the herd effect.	

Mark scheme for Paper D

Section 1 (20 marks – 7 PS, 13 KU)

Question	Type	Mark	Expected response	Comment
1	KU	1	D	Be familiar with the differences between somatic and germline cells and the genetic outcomes of mitosis and meiosis.
2	PS	1	C	First, work out the mass of urea in 1L of blood as $1 / 60 \times 6 = 0{\cdot}1$ g. $100\ cm^3$ is $1 / 10$ of a litre so the final calculation is $1 / 10 \times 0{\cdot}1 = 0{\cdot}01$ g.
3	KU	1	A	Insulin is formed by the combination of two polypeptide chains after translation.
4	KU	1	B	
5	PS	1	D	Be careful not to go beyond what is given here. The fact that drug 2 dissolves fastest does not allow you to conclude it will be the most effective at treating pain. Drug 6 actually takes longer than drug 5 to dissolve. There is no data to support or refute how well these drugs dissolve in water. Drug 2 takes the least time to dissolve.
6	KU	1	A	
7	KU	1	D	
8	KU	1	D	The hormones involved in human reproduction are varied and complex in their interaction. Construct good diagrams to help remember their names and how they influence each other.
9	PS	1	B	Do not be put off by this table which looks a little complicated. It is typical of data used to show trends over a period of time and also make comparisons between different groups, here males and females. Survival rates for men with cancer of the oesophagus have increased from 12% to 22%, an overall increase of 10%. For stomach cancer, the increase has been from 13% to 25%, an overall increase of 12%, greater than that of cancer of the oesophagus so A is true. Over the whole period, comparing women with men for survival of stomach cancer, women have always fared better so B is false. Between 1986 and 1990, 48% of people (males and females together) survived cancer of the oesophagus whereas 52% of people (males and females together) survived stomach cancer so C is true. Between the periods 1976–1980 and 1981–1985, in men the survival increased from 14% to 20%, an overall increase of 6% which was the greatest increase so D is true.

10	KU	1	D	Notice the father is unaffected and the mother is a carrier. If the allele for this condition was dominant the mother could not be a carrier, whether or not the condition was sex-linked. We can conclude therefore the allele must be recessive. The father is unaffected and therefore must have the normal dominant allele while the mother must be heterozygous since she is free from any symptoms but is a 'carrier'. By studying the pattern of behaviour in their four children, it is possible to see the condition must be sex-linked since they have an affected son. Remember, the affected son inherits one of each form of allele from each parent. To be 'affected' his father would have to be able to pass on the defective allele but he can't since he has the dominant normal form. We can be sure, therefore, the condition is both sex-linked and caused by a recessive allele.
11	KU	1	A	Tissue fluid which is not reabsorbed into the blood ends up in blind-ending tubes called lymphatic vessels.
12	PS	1	A	
13	KU	1	B	
14	PS	1	B	From the diagram, you can see that people over 85 years of age are predominantly female. Over 300,000 males are in the range 45–49 years, the largest of any of the given ranges. This is therefore a valid conclusion. There are far more females older than 20 years in this sample. In the oldest range, there are fewer males than females.
15	KU	1	A	
16	PS	1	C	As we get older, our ability to regulate our body temperature becomes less efficient. An older person also has a much larger surface area to volume ratio. These two variables combined cause an older person to be at much greater risk of hypothermia than a younger person.
17	PS	1	D	From 120 to 150 minutes, the 18-year-old's body temperature dropped from 36·4°C to 36·0°C, a drop of 0·4°C. If this continued over the next 30 minutes, his new temperature would be 36·0 – 0·4 = 35·6°C.
18	KU	1	A	
19	KU	1	A	Remember verbal communication can be spoken and/or written.
20	KU	1	D	

Section 2 (80 marks – 27 PS, 53 KU)

Question		Type	Mark	Expected response	Comment
1	a	KU	1	DNA polymerase.	
	b	KU	2	Strand 1 correctly labelled as new one moving toward the replication fork. Strand 2 correctly labelled as one moving away from replication fork.	
	c	KU	1	DNA is unwinding and unzipping.	
2	a	PS	1	Five hours.	After five hours, the concentration in the mucus with the drug is 1·6 µg/ cm^3 compared with no drug at 0·2 µg/cm^3. This is a difference of eight times in the presence of the drug.
	b	PS	2	With the drug, the concentrations in all the mucus samples are always higher than without the drug. With the drug, the concentration in the mucus samples reaches a much higher level than without the drug. With the drug, after one hour, the concentration in the mucus sample is much higher than without the drug. With the drug, the concentration in the mucus samples stays at a higher level than without the drug.	There are several other differences which could be given but notice the precise use of the language in answering. It is possible to state the same differences in reverse by starting with 'Without the drug …'. Any two valid differences, such as are given here, would each get one mark.
	c	PS	2	Sample size was very small. No indication of gender. No indication of level of infection. One group is given an additional drug but the control group was not given a mock/placebo type drug.	Again, there are several other design features which would invalidate a proper conclusion. Any two valid features, such as are given here, would each get one mark.
	d	PS	3	Correct scales and labels on axes. Plotting points correctly and drawing lines. Labelling/unique two lines.	Practise drawing different kinds of graph so you pattern the correct method of labelling the axes, joining the points, adding units etc.
	e	PS	1	The use of the additional drug raises the level of antibiotic in the mucus samples.	As before, it is acceptable to turn this around and say 'Without the use of the drug …'

3	a		KU	1	Generate billions of copies of a specific DNA sequence.	
	b		KU	2	Stage X Stage Z Stage Y	All three correct – 2 marks. One or two correct – 1 mark.
	c		KU	1	54°C to 72°C = 18°C	
	d		KU	1	Allow DNA polymerase to bind.	
	e		PS	1	12,800	Each cycle lasts 5 minutes so after 35 minutes, 7 cycles would occur with the quantity of DNA doubling each time.
	f		KU	2	Generate DNA profile/genetic fingerprint. Research into conditions caused by mutations. Test for a known genetic condition. In archaeology to make links between finds and other living/non-living relations. Classification studies to map how species have evolved.	There are many examples of how PCR can be applied. Make sure you know a range of examples.
4	a	i	KU	1	Ovulation.	
		ii	KU	1	Follicle-stimulating hormone/FSH.	It is acceptable to use the standard abbreviations for words like follicle-stimulating hormone, deoxyribonucleic acid, adenosine triphosphate and so on. However, you should also know the full name of each.
		iii	KU	1	Pituitary.	The pituitary has been likened to the conductor of an orchestra and is often nicknamed the 'master gland' because it co-ordinates so many different activities. It produces many hormones which have direct effects on other organs as well as indirect effects by stimulating endocrine glands to secrete their own hormones. You would do well to construct a simple diagram of the pituitary with the various hormones and their effects shown.

		iv	KU	1	Corpus luteum.	
		v	KU	2	Progesterone. Oestrogen.	
	b		KU	1	During *in vitro* fertilisation, the egg and sperm are allowed to mix outside the body of the female.	*In vitro* literally means in glass and is an apt description of how this process is artificial, outwith the female's body. It is often useful to have an understanding of Latin terms used in Biology, rather than learning by rote.
5	a	i	KU	1	Huntington's disease.	Notice that males and females are affected by this disease so it is not sex-linked. Also, the disease is expressed in relatively high numbers which means it is unlikely to be recessive. This leads, by elimination, if you were not able to work it out from first principles, to the conclusion that the disease is autosomal dominant and therefore Huntington's.
		ii	PS	2	Condition appears in each generation. Each affected person has an affected parent. If both parents are unaffected, the condition does not appear in the next generation.	
	b		PS	1	50%	Suppose you represent the allele for Huntington's disease by H, then A's father is Hh (since his daughter is unaffected) and his mother is hh. Therefore there is a 50% chance he will inherit the allele causing the disease.
	c		PS	1	250	Half of 5,000,000 is 2,500,000 and when you divide this by 10,000 you get the answer of 250.
	d		KU	1	Chorionic villus sampling.	

6	a	i	PS	2	Maximum heart rate stays higher longer than fertility. Fertility stays at 100% for longer than maximum heart rate. Fertility falls rapidly after 25 years compared with maximum heart rate which falls at a steady, slower rate. Fertility reaches a 0% value whereas maximum heart rate does not. Fertility falls faster than maximum heart rate from ages 25 to 50 years.	There are a variety of valid conclusions from this data. Any two of these would each get one mark.
		ii	PS	1	65 beats per minute.	Every ten years the maximum heart rate falls by approximately 5 beats per minute. At 70 years it is 70 beats per minute so at 80 years it will be about 65 beats per minute.
		iii	PS	1	15 and 20 years.	The fertility rises by 65%, the greatest change for any 5-year period.
		iv	PS	1	5 years.	Between the ages of 20 and 25 years.
	b		PS	1	4	25% of 200 = 50 women aged 40 years or more. Of this 50, 8% will experience menopause. $8/100 \times 50 = 4$.

| 7 | a | | PS | 2 | There is no indication of whether the subjects all were originally caffeine drinkers. The poorer performance of the people given no caffeine could have been due to the effect of not having any caffeine. The people on the caffeine drink could simply be performing at their 'normal' level and not necessarily be showing an improvement. It may take longer than 24 hours for caffeine to leave the body. | If a person does/does not intake caffeine regularly, the effect of giving them a caffeine drink could well be very different. Ideally, all the subjects should be either regular partakers or regular abstainers. It is well known that people who take caffeine regularly and then have this withdrawn have lower performance due to poor alertness, becoming tired and so forth. The people on the non-caffeine drink could be performing badly for this reason. There is no way of knowing if the 'enhanced' performance of the caffeine drinkers is any better or any worse than what they would normally achieve. It is known that it can take much longer than 24 hours for caffeine to leave the body completely. There are other possible reasons. Any two of these given or reasonable alternatives would each obtain one mark. |

	b		PS	2	Ensure the subjects were all regular caffeine drinkers/abstainers. Do not divide the subjects into two groups but devise a method for tagging what they drink so they are unaware of being in one or other group. Make sure the experimenters are unaware of what each drink contains.	Be careful here not to duplicate what was said in (a) above. Make sure to give alternative changes in the experimental design which don't use the answers already given. It is an improvement on the design of the experiment not to 'allocate' people into groups, even if the person doesn't know whether or not they are receiving a caffeine or non-caffeine drink. These types of experiments are always performed 'double-blind' with both the subjects and the experimenters, in this case, being unaware of what is in the drink, in this case. There are other design improvements and so long as you don't duplicate anything, you would get one mark for each correct suggestion.
8	a		KU	2	Peripheral nervous system. Somatic nervous system.	
	b		KU	2	Sympathetic: Usually speeds up functions. Prepares body to deal with stress. Sets up 'flight or fight' response. Parasympathetic: Usually slows down functions. Increases rate of digestion. Decreases breathing rate and depth.	There are a number of possible answers here.
	c		KU	1	Corpus callosum.	
	d		KU	1	Processing information associated with memory formation. Important in long-term memory formation. Influencing emotional state.	There are a number of possible answers here.
9	a		KU	2	Released from pathways involved in euphoria. Release increases during stress.	

	b		KU	1	To treat Parkinson's disease.	
	c		KU	2	By action of specific enzymes. By being taken back up by pre-synaptic membrane.	
	d		KU	1	Mimics the normal neurotransmitter.	
10	a		PS	1	Instead of learning by rote, if we form a mental image which sounds like the word we are trying to learn.	
	b		PS	1	Allow subjects to make up their own images and keywords.	
	c		PS	1	The method will not work if the foreign word does not have a direct word in English.	This is quite a difficult concept to grasp. Only if the foreign word can be translated directly into a word in English will this strategy work as illustrated here.
11	a	i	KU	2	Immune system is stimulated to produce antibodies on first exposure to antigen A, reaching peak after just over three weeks. On meeting the same antigen A again, the immune system responds much quicker and to a higher level. [either of these points for 1 mark] This is due to immunological memory. [required for second mark]	Make sure you understand the differences between primary and secondary immune responses to the same antigen.
		ii	KU	2	Exposure to antigen B has not taken place until week seven so this is a primary response. Antibody concentration changes mirror those for first exposure to antigen A. [either of these points for 1 mark] Immunological memory will be formed for a secondary exposure to the same antigen/no immunological memory has been previously formed. [required for second mark]	
	b		KU	1	Rheumatoid arthritis/multiple sclerosis/type 1 diabetes	

12	a		KU	3	Quarantine Isolate individual who has or has been exposed to an infectious disease. Maintain isolation over the normal incubation period of infection. Prevents spread to non-infected people. Antisepsis Protocols which kill or inhibit the growth of potential pathogens. Include measures such as sterilisation of equipment, chemical disinfectants, wearing of masks and/or gloves. Antiseptic chemicals can be applied safely to living tissue, killing or inhibiting potential pathogens but not damaging host. Vector control Some infections are spread by organisms such as mosquitoes/ticks to other animals. For example, malaria by being bitten by infected mosquito or Lyme disease via infected tick. Reducing/eliminating population of vector will reduce/eliminate spread of associated infectious agent.	
	b		KU	2	Sporadic 1 Epidemic 3 Endemic 2	All three correct – 2 marks. One or two correct – 1 mark.
	c		KU	1	Good personal hygiene/hand-washing/use of antibacterial soaps and gels/using condoms associated with sexual activity/observing good protocols in handling and/or cooking and/or storing food.	
13	A	i	KU	5	1. Failure to ovulate; use of fertility drugs to treat. 2. Ova viability low; use donated ova. 3. Oviducts blocked; surgery to remove blockage/*in vitro* fertilisation. 4. Failure of embryo to implant; treat with fertility drugs. 5. Hormone imbalances leading to poor ovulation; use ovulation-stimulating drugs and/or *in vitro* fertilisation.	

		ii	KU	5	1. Physical methods rely on some kind of barrier preventing sperm fertilising ovum or successful implantation. 2. Condoms stop sperm entering vagina. 3. Vasectomy prevents sperm travelling along sperm duct. 4. Diaphragm inserted into female reproductive tract prevents sperm meeting ova. 5. Intra-uterine device makes implantation difficult should fertilisation occur.	
	B	i	KU	5	1. Non-specific defence is usually body's first line response to potential infection. 2. Responds quickly but is not long-lasting. 3. Skin and epithelial cells form primary physical barrier preventing penetration of infectious agents. 4. These cells also secrete chemicals such as mucus, tears, sweat, saliva, wax which trap/immobilise/kill potential pathogens. 5. Some epithelial cells have cilia which move potential pathogens out of the respiratory system. 6. Low pH of stomach kills most microorganisms which have been ingested. 7. Secretion of cytokines helps organise phagocytes to migrate to sites of potential infection.	There are other examples of non-specific defence mechanisms. Make sure you give at least 5 different points in your answer.

		ii	KU	5	1. Clinical trials must yield objective evidence of safety, effectiveness and valid results.	Make sure you give at least 5 different points in your answer.
					2. Testing on human volunteers usually involves setting up a test/experimental group and a control group.	
					3. Volunteers in test group are given vaccine and those in control group are given a placebo.	
					4. No volunteer knows which group he/she belongs to.	
					5. The experimenters are unaware of which group a volunteer belongs to and whether the vaccine they are administering is real or a placebo.	
					6. Double-blind protocol.	
					7. Reliability is increased by a large sample size making results statistically more significant.	

© 2017 Leckie & Leckie Ltd
Cover image © ink-tank

001/26042017

10 9 8 7 6 5 4 3 2 1

ISBN 9780008209018

Published by
Leckie & Leckie Ltd
An imprint of HarperCollins*Publishers*
Westerhill Road, Bishopbriggs, Glasgow,
G64 2QT
T: 0844 576 8126   F: 0844 576 8131
leckieandleckie@harpercollins.co.uk
www.leckieandleckie.co.uk

Commissioning Editor: Clare Souza
Managing Editor: Craig Balfour

Special thanks to
Claire Coombe-Jones (copyedit)
Jess White (proofread)
Louise Robb (proofread)
QBS (layout and illustration)

Printed in Italy by Grafica Veneta S.p.A

A CIP Catalogue record for this book is
available from the British Library.

Acknowledgements
P90 Oksana Kuzmina/Shutterstock.com

MIX
Paper from
responsible sources
FSC™ C007454

FSC™ is a non-profit international organisation established to promote
the responsible management of the world's forests. Products carrying the
FSC label are independently certified to assure consumers that they come
from forests that are managed to meet the social, economic and
ecological needs of present and future generations,
and other controlled sources.

Find out more about HarperCollins and the environment at
www.harpercollins.co.uk/green